Measuring the Strategic Value of the Armed Forces Health Longitudinal Technology Application (AHLTA)

T0195526

James H. Bigelow, Katherine M. Harris, Richard Hillestad

Prepared for the Office of the Secretary of Defense

Approved for public release; distribution unlimited

Center for Military Health Policy Research

A JOINT ENDEAVOR OF RAND HEALTH AND THE
RAND NATIONAL DEFENSE RESEARCH INSTITUTE

The research reported here was sponsored by the Office of the Secretary of Defense (OSD). The research was conducted jointly by the Center for Military Health Policy Research, a RAND Health program, and the Forces and Resources Policy Center, a RAND National Defense Research Institute (NDRI) program. NDRI is a federally funded research and development center sponsored by the OSD, the Joint Staff, the Unified Combatant Commands, the Department of the Navy, the Marine Corps, the defense agencies, and the defense Intelligence Community under Contract W74V8H-06-C-0002.

Library of Congress Cataloging-in-Publication Data is available for this publication.

ISBN 978-0-8330-4314-6

The RAND Corporation is a nonprofit research organization providing objective analysis and effective solutions that address the challenges facing the public and private sectors around the world. RAND's publications do not necessarily reflect the opinions of its research clients and sponsors.

RAND® is a registered trademark.

Published 2008 by the RAND Corporation
1776 Main Street, P.O. Box 2138, Santa Monica, CA 90407-2138
1200 South Hayes Street, Arlington, VA 22202-5050
4570 Fifth Avenue, Suite 600, Pittsburgh, PA 15213-2665
RAND URL: http://www.rand.org/
To order RAND documents or to obtain additional information, contact
Distribution Services: Telephone: (310) 451-7002;
Fax: (310) 451-6915; Email: order@rand.org

Preface

The Military Health System (MHS) has more than nine million eligible beneficiaries, including active duty service members and their families, retirees and their families, and Guard and Reserve members serving on active duty and their families. The MHS provides health care through its own facilities and personnel (direct care); it also purchases care from civilian providers (purchased care). In January 2004, the MHS's Clinical Information Technology Program Office (CITPO) began implementation of the Armed Forces Health Longitudinal Technology Application (AHLTA), DoD's global electronic health record. AHLTA will ultimately be used by all providers in the military's direct care system at the point of care. It will also promote population health, conduct medical surveillance, support clinical decisionmaking, and support force health protection for deployed service members. As of December 2006, AHLTA was being used to document virtually all outpatient care delivered at fixed MHS facilities.

We undertook this project between June 2006 and October 2007 at the request of the Assistant Secretary of Defense for Health Affairs (OASD/HA). Our objective was to help MHS develop an analytic framework and define specific outcome measures for assessing and reporting the efficiency, safety, and health benefits of AHLTA as it becomes fully deployed.

This monograph describes the framework we recommend that DoD adopt in measuring AHLTA's contribution to MHS performance. To develop the framework, we (1) reviewed AHLTA's current and planned capabilities, (2) reviewed the literature on the measured benefits of health information technology, (3) consulted with senior MHS leaders to understand the dimensions of performance that the leadership deemed important and how the leadership anticipated that AHLTA would affect those dimensions, (4) identified and assessed performance measures in current use by civilian health system for their applicability to MHS strategic objectives, and (5) suggested new approaches for measuring MHS strategic objectives where civilian measures are lacking.

This study was conducted jointly by RAND Health's Center for Military Health Policy Research and the Forces and Resources Policy Center of the National Defense Research Institute (NDRI). NDRI is a federally funded research and development center sponsored by the Office of the Secretary of Defense, the Joint Staff, the Unified Combatant Commands, the Department of the Navy, the Marine Corps, the defense agencies, and the defense Intelligence Community.

For more information on RAND's Forces and Resources Policy Center, contact the director, James Hosek. He can be reached by email at james_hosek@rand.org; by phone at 310-393-0411, extension 7183; or by mail at the RAND Corporation, 1776 Main Street, Santa Monica, California 90407-2138. Susan Hosek and Terri Tanielian are co-directors of the RAND Center for Military Health Policy Research. Susan Hosek may be reached by email at sue@rand.org; by phone at 310-393-0411, extension 7255; or by mail at RAND Corporation, 1700 Main Street, Santa Monica, California 90407. Terri Tanielian may be reached by email at territ@rand.org; by phone at 703-413-1100, extension 5265; or by mail at RAND Corporation, 1200 South Hayes Street, Arlington, Virginia 22202. More information about RAND is available at www.rand.org.

Contents

Figure and Tables

Summary

Background

The purpose of this project, which was undertaken between June 2006 and October 2007, was to help the Military Health System develop an analytic framework and define specific outcome measures for assessing and reporting the efficiency, safety, and health benefits of AHLTA, DoD's electronic medical record system, as it becomes fully deployed.

Deployment of AHLTA is occurring in several planned phases or "blocks." Block 1 deployment to military treatment facilities (MTFs) throughout the world began in January 2004. The timing of Block 4 deployment (the last planned block) has yet to be determined. As of the end of 2006, AHLTA Block 1 had been installed at 138 fixed military installations worldwide, and virtually all outpatient care they delivered was being documented in AHLTA's clinical data repository (CDR). The CDR can be accessed (with appropriate permissions) from any installation, so if a beneficiary is treated at one installation and later seeks care at another installation, his record is available.

The CDR cannot now be accessed electronically from a theater of operations, such as Iraq or Afghanistan.[1] The MHS is addressing this problem through the Theater Medical Information Program-Joint (TMIP-J).

AHLTA will gain functionality as blocks 2–4 are rolled out. By the end of 2007, Block 2 will add ordering and management of eye-

---

[1] AHLTA has been deployed at a number of permanent overseas installations in Germany, South Korea, and Japan. The CDR can also be accessed from these locations.

glasses and dental charting and documentation. By the end of 2011, Block 3 is scheduled to replace legacy pharmacy, laboratory, anatomical pathology, radiology, occupational health, and surveillance capabilities. Also by the end of 2011, Block 4 will extend AHLTA to inpatient encounters. By this time, essentially all aspects of health care provided by the MHS (but not care purchased from civilian providers) will be documented in AHLTA.

The MHS is simultaneously deploying a clinical data mart (CDM), which imports data from AHLTA's CDR and facilitates enterprise-level analysis and decision support. The CDM was in operation as of the end of 2006, but its analysis capability was immature. There are also plans to implement a clinical data warehouse (CDW). This capability would access and link data from sources beyond the CDR—such as eligibility and enrollment data, accounting data, and surveys—and would have the potential to support more complex analyses than does the CDM.

Views of Senior MHS Leadership

Senior leaders of the MHS will be the primary customers for measures of the performance of the MHS and of the strategic value of AHLTA. Therefore, we conducted a series of 14 interviews with senior MHS leaders. Through these interviews we discovered the dimensions of performance that the leadership deemed important, and how the leadership anticipated that AHLTA would affect those dimensions. They were unanimous that the correct dimensions of performance to use were the strategic objectives contained in their strategic plan (OASD, 2007a). The strategic objectives go beyond efficiency, safety, and health to include the medical readiness of service members to deploy and the ability of the military medical system to provide outstanding health care in support of military operations.[2]

The leadership anticipated that AHLTA will have effects on performance similar to those reflected in the published literature on electronic medical records (EMRs). These effects include improved qual-

---

[2] Specific strategic objectives are depicted in Figure 2.1.

ity and safety of care through point-of-care alerts and reminders and reduced cost through avoidance of duplicate tests. AHLTA-generated reminders, they expected, will improve individual medical readiness. Improved continuity of care will benefit service members evacuated from theater with wounds, injuries, or disease. A few participants mentioned AHLTA's potential as a tool to support process improvement efforts through measurement and feedback.

The Framework

To assess the strategic value of AHLTA, we recommend the standard treatment-outcomes[3] methodology described in textbooks on program evaluation (Mohr, 1988; Rossi, Freeman, and Lipsey, 1999). The methodology has four elements:

1. *Outcome measures* capturing valued domains of system performance plausibly influenced by the presence of AHLTA
2. A *treatment* describing a way of using AHLTA that is expected to influence one or more outcome measures
3. A *logic model* describing the mechanisms or processes by which uses of AHLTA (i.e., treatments) influence outcome measures
4. An *evaluation design* for estimating AHLTA's effect on outcome measures in quantitative terms.

Given AHLTA's complexity and evolving nature, we do not consider AHLTA, as a whole, to be a "treatment." Instead, we consider a treatment to be a specific use of AHLTA under defined circumstances. An example of a treatment is the following: "When an active duty service member arrives for an appointment, determine whether the patient meets all the criteria to be medically ready to deploy; if not, remind the physician to inform the patient of his readiness status."

---

[3] Throughout this monograph we use the word *treatment* as it is used in the program evaluation literature: the action or intervention whose effects on outcomes is to be estimated. This is more general than the use made of the word in healthcare, where a treatment is something done to a patient to affect his health status.

Logic models trace the cause-and-effect chains through which treatments influence outcome measures. Each treatment requires its own logic model, although many treatments should have very similar models. A logic model describes cause-and-effect relationships qualitatively, and data are needed to turn the qualitative arguments into quantitative estimates.

An evaluation design is the strategy for separating a treatment's quantitative effect on outcome measures from the confounding effects of other factors. The evaluation design will determine what data are needed and how those data could be used to generate the estimates. It is standard practice to measure this effect by comparing outcomes observed in the presence of the treatment to estimates of the outcomes that would have occurred in the absence of the treatment.

If the data are insufficient to estimate AHLTA's effects quantitatively, it is possible to use methods that combine logic models and a mixture of qualitative and quantitative evidence to support the existence of a cause-and-effect relationship between AHLTA and relevant outcomes. However, such methods rely heavily on quantitative measures of AHLTA-relevant outcome measures.

Specific Measures

The MHS strategic objectives are too abstract to serve as outcome measures. We have suggested detailed measures (i.e., measures specific to care setting, condition, procedure, and patient characteristics) for a wide range of strategic objectives that AHLTA could plausibly influence. Where possible, we suggest using detailed measures of healthcare quality that are in the public domain, which can be obtained from the National Quality Measures Clearinghouse (NQMC). While the NQMC measures represent the practical universe of existing quality measures, they cover only a minority of MHS strategic objectives. Measures for other strategic objectives must be developed from scratch. Where possible, we have suggested what those measures might be.

Before they are used, however, measures should be assessed for *reliability* (the degree to which the measure is free from random variation) and *validity* (the degree to which the measure is associated with what it purports to measure).

In order for high level managers to avoid information overload, we discuss methods for creating higher-level measures by selecting, aggregating, and summarizing the detailed measures. With this approach, it remains possible to drill back down to more-detailed levels to pinpoint a problem when a high-level measure reveals a problem.

Implementation Issues

Calculating a detailed outcome measure requires specific data elements and a precisely defined algorithm for combining them. To implement a given measure, therefore, one must identify sources for each data element, and those sources must reliably contain correct values for the data elements. Moreover, the data should not be buried in free text. For the measure to be practical, it must be possible to retrieve the needed data automatically.

Many measures of potential interest cannot be implemented because the necessary data are not collected. If the MHS deems a measure sufficiently important, MHS leadership may choose to change their information systems and policies to collect the necessary data. AHLTA enhancements are fielded regularly, and the need to collect additional data may be the purpose of future enhancements. Changes for this purpose could be made to other information systems; they need not be confined to AHLTA alone.

Finally, to implement our proposed framework, there must be an organizational home for a measurement system. The organization chosen will need ready access to data from many other systems as well. In addition, it will need considerable analytic capability. We anticipate that many of the treatments to be analyzed will originate as process improvement exercises at individual installations. That organization will need to provide analytic support for these efforts, and it should act as a clearinghouse for information on all installations' improvement efforts.

Acknowledgments

We wish to thank Carl Hendricks, Connie Gladding, Sharon Larson, and Col James Benge for their assistance and persistence in arranging interviews with senior MHS officials. We are grateful to Michael Dinneen and Paul Tibbits for sharing their insights on the role of health information technology in the strategic transformation of the Military Health System. We are grateful for the assistance of Andrew Baxter and Qiufei Ma for their help in surveying existing sources of quality measures. We also wish to thank the officials who participated in our interviews and the individuals who hosted our site visit and provided a demonstration of AHLTA's capabilities. Michael Hix and Robin Meili provided valuable comments on an earlier draft of this report. Their suggestions have greatly improved the final product.

Abbreviations

AHLTA	Armed Forces Health Longitudinal Technology Application
AHRQ	Agency for Healthcare Research and Quality
BSC	balanced scorecard
CCM	Chronic Care Model
CDM	clinical data mart
CDR	clinical data repository
CDW	clinical data warehouse
CHCS	Composite Health Care System
CITPO	Clinical Information Technology Program Office
CMS	Centers for Medicare and Medicaid Services
CMS	Command Management System
COPD	chronic obstructive pulmonary disease
CPOE	computerized physician order entry
DEERS	Defense Enrollment Eligibility Reporting System
DMHRSi	Defense Medical Human Resource System-Internet
EIDS	Executive Information Decision Support
EMR	electronic medical record
ESSENCE	Electronic Surveillance System for the Early Notification of Community-based Epidemics
ETG	Episode Treatment Group
HA	Health Affairs

HEDIS	Health Plan Employer Data and Information Set
HIP	health innovations program
HIT	health information technology
HRSA	Health Resources and Services Administration
ICSI	Institute for Clinical Systems Improvement
IM/IT	information management/information technology
IMR	individual medical readiness
LDL-C	low density lipoprotein – cholesterol
MEB	Management Evaluation Board
MEDPROS	Medical Protection System
MEG	Medical Episode Grouper
MEPRS	Medical Expense and Performance Reporting System
MHS	Military Health System
MTF	military treatment facility
NCQA	National Committee for Quality Assurance
NGC	National Guideline Clearinghouse
NHQR	National Healthcare Quality Report
NQMC	National Quality Measures Clearinghouse
OASD	Office of the Assistant Secretary of Defense
OIG	Office of the Inspector General
OMB	Office of Management and Budget
PDTS	Pharmacy Data Transaction Service
PMPM	per member per month
RVU	relative value units
RWP	relative weighted product
SADR	Standard Ambulatory Data Record
SIDR	Standard Inpatient Data Record
TMA	TRICARE Management Activity
TMIP-J	Theater Medical Information Program–Joint
VHA	Veterans Health Administration

CHAPTER ONE

Introduction

Objective and Rationale

The purpose of this project, which was undertaken between June 2006 and October 2007, was to assist the Military Health System (MHS) in developing an analytic framework and in defining specific outcome measures for assessing and reporting the efficiency, safety, and health benefits of the Armed Forces Health Longitudinal Technology Application (AHLTA), DoD's electronic medical record system, as it becomes fully deployed.

Federal law dictates how federal investments in information technology will be assessed. The Information Technology Management Reform ("Clinger-Cohen") Act of 1996 (P.L. 104-106) orders the Director of the White House Office of Management and Budget to

- Periodically review information technology investments to determine its "efficiency and effectiveness [...] in improving the performance of the executive agency and the accomplishment of the missions of the executive agency."
- Ensure "that the performance measurements measure how well the information technology supports programs of the executive agency;" and
- Use "comparable processes and organizations in the public or private sectors exist, quantitatively benchmark agency process performance against such processes in terms of cost, speed, productivity, and quality of outputs and outcomes."

The Clinger-Cohen Act and guidance[1] issued by the Office of Management and Budget (OMB) to assist agencies in implementing the Act are clear that return on federal investments in information technology, such as AHLTA, are to be assessed in terms of their effect on high-level measures of agency performance. To meet these requirements, the MHS must develop a system for measuring AHLTA's contribution to the strategic performance of the MHS in terms of its effect on efficiency, safety, readiness, and health outcomes.

Background of the Military Health System

Roughly nine million active duty service members, active duty family members, and retirees are eligible to receive medical care through the MHS. Eligible beneficiaries receive MHS-sponsored care through three health plans offered by TRICARE: a plan similar to a health maintenance organization called "TRICARE Prime" that delivers care primarily through MTFs, a preferred provider network called "TRICARE Extra," and a fee-for-service plan called "TRICARE Standard" (TRICARE 2007, p. 6).[2] Active duty service members are automatically enrolled in TRICARE Prime. Other eligible beneficiaries may choose among the three options.

The MHS provides two types of care to its beneficiaries. It provides *direct care* through military hospitals and clinics, commonly called military treatment facilities (MTFs) and *purchased care* provided by civilian providers financed through managed care contracts and fee-for-service reimbursements. Within the direct care system, each military service, under its surgeon general, is responsible for managing its MTFs. In 2006, the direct care system provided less than 40 percent of all inpatient, outpatient, and pharmacy services used by MHS benefi-

---

[1] For example, OMB, *Capital Programming Guide, Supplement to Part 7 of Circular No. A-11.*

[2] Additional health plan options are available to certain National Guard and Reserve members and their families and to Medicare-eligible MHS beneficiaries living in regions where TRICARE Prime is not available (TRICARE, 2007, pp. 16, 32).

ciaries and accounted for roughly half of all MHS medical care expenditures (Tables 1.1 and 1.2).

Under current plans, AHLTA will be implemented only in the direct care system. The direct care system provides care to three distinct groups of MHS beneficiaries: (1) 1.70 million active duty service members who are automatically enrolled in TRICARE Prime, (2) 5.15 million eligible beneficiaries who live in defined catchment areas surrounding MTFs and have elected to enroll in TRICARE Prime, and (3) the remaining 2.6 million eligible beneficiaries permitted to use MTF care on a space-available basis (TRICARE 2007, pp. 15, 16).

The two groups of TRICARE Prime enrollees use MTFs as their primary source of care and can be referred to purchased care providers in situations where it is clinically appropriate or access to MTF care is constrained. Together, active duty service members and TRICARE

Table 1.1
Number of Events in Direct and Purchased Care Systems, FY 2006

Venue	Direct	Purchased	Total
Inpatient stays (thousands)	263.2	805.8	1,069.0
Outpatient encounters (millions)	31.2	53.3	84.5
Outpatient drugs (millions of scripts)	49.1	66.6	115.7

SOURCE: TRICARE, 2007, pp. 20, 21.

Table 1.2
Expenditures on Direct and Purchased Care, FY 2006 ($ millions)

Venue	Direct	Purchased	Total
Inpatient stays	$2,587	$3,317	$5,904
Outpatient encounters	6,365	5,770	12,135
Outpatient drugs	2,645	4,940	7,585
Total	11,597	14,027	25,624

SOURCE: TRICARE, 2007, pp. 22, 23.

Prime members constitute roughly the 75 percent of MHS beneficiaries whose care use of direct care is documented in AHLTA.

AHLTA Background

AHLTA will provide real-time longitudinal health records that are accessible 24 hours a day, seven days a week at military health installations around the world. The $5 billion system[3] began phase-in across the force in January 2004, and deployment is expected to be complete by the end of 2011. As of the end of 2006, virtually all outpatient care delivered at fixed facilities worldwide was being documented in AHLTA.

Deployment of AHLTA is occurring in several planned phases or "blocks." Block 1 deployment was completed in December 2006 (MHS Conference, 2007a), though enhancements will continue to be rolled out as the system is used. Block 1 functions include

- encounter documentation
- order entry and results retrieval
- encounter coding support
- consultation tracking
- alerts and reminders, including some automated clinical practice guidelines
- role-based security—i.e., the data and functions that users can access depend on their roles in the system
- health data security
- master patient index.

The encounter coding support function is based on a feature of AHLTA that is rare among electronic medical records (EMRs)—the requirement to use structured terms to document each patient encounter. In the vast majority of EMRs (e.g., the Veterans Administration's

---

[3]   This is the life cycle cost as reported by the DoD Office of the Inspector General (OIG, 2006), which includes 17 years of operating costs.

VISTA and Kaiser Permanente's EMR from Epic Systems), the physician chooses the terms used to document a patient encounter, and two physicians may use different terms to document similar observations. AHLTA, by contrast, requires physicians to use a standardized terminology to document their observations, although they retain the option to enter explanatory free text.[4] The advantage of using structured terms is that they can be processed by computer programs much more extensively than free text can. One use of this processing is encounter coding support. Another use is public health surveillance.

Block 1 documents only outpatient encounters (not inpatient stays), and only if they occur at one of the 138 permanent installations where AHLTA has been installed.[5] Medical record entries generated at any installation are stored in a clinical data repository (CDR) that can be accessed (with appropriate permissions) from any other installation. Thus, if a beneficiary is treated at one installation and later seeks care at another installation, his record is available.

The CDR receives medical information from a theater of operations through the Theater Medical Information Program-Joint (TMIP-J).[6] The first responder captures clinical data on AHLTA Mobile, a handheld device that stores medical data until there is an opportunity to download it to AHLTA Theater. AHLTA Theater operates on a laptop, and in turn stores medical data until communications are available to send the data to the CDR. Previous AHLTA encounters can be viewed from the theater shortly after they are signed through a Web-based application called the Theater Medical Data Server, which is accessed through the Medical Communications for Combat Casualty Care system.

---

[4] In principle, a physician can use the free-text option exclusively, thus undermining the ability for structured and quick analyses. In practice, a considerable amount of structured documentation is done.

[5] Since the 1980s, the MHS has documented inpatient stays at its direct care hospitals in an electronic record called the Composite Health Care System (CHCS I). (AHLTA used to be called CHCS II.) CHCS I records are kept locally and cannot be shared electronically between MTFs. However, AHLTA can access the local inpatient record. As mentioned later, documentation of inpatient stays will migrate to AHLTA with the deployment of Block 4.

[6] See the TMIP-J Web site for a description of the program.

AHLTA will gain functionality as blocks 2–4 are rolled out. By the end of 2007, Block 2 will add ordering and management of eyeglasses and dental charting and documentation.

By the end of 2011, Block 3 will have replaced legacy systems that perform the following ancillary functions:

- pharmacy
- laboratory
- anatomical pathology
- radiology
- occupational health and surveillance.

Eventually, Block 4 will extend AHLTA to inpatient encounters. By this time, AHLTA will document all aspects of direct care delivered by the MHS. Purchased care will only be documented in AHLTA to the extent that MHS health care providers scan records of purchased care into AHLTA. Information from scanned records will be in text or image form and thus not computable or analyzable. Likewise, providers of purchased care will not have access to patient records through AHLTA.

The documentation of all encounters at facilities using AHLTA flows into the CDR. Data in the CDR are organized in such a way as to make it easy to access the complete record of any individual patient. The data in the CDR are copied into a clinical data mart (CDM) where multiple patient records can be more easily manipulated and organized to support enterprise-level analysis and decision support. The CDM was in operation as of the end of 2006, but its capabilities to support data analysis were immature. There are also plans to implement a clinical data warehouse (CDW). This capability would support analysis of data sources beyond the CDR (e.g., eligibility and enrollment data, accounting data, and surveys) and would have the potential to support more complex analyses than the CDM.

Implementing AHLTA will create the potential for benefits of the kinds documented in a two-year RAND research project of EMR systems (Hillestad et al., 2005). In this study, the EMR system was shown to have the potential to reduce duplicative testing, reduce adverse drug events, increase preventive care, increase adherence to evidence-based practice, and improve the care of chronic diseases. These benefits stem from the fact that the EMR gives clinicians access to more information more quickly. But implementing an EMR system does not assure these benefits. Clinicians must make good use of the extra information to realize the benefits, and that means they must change the way they do their work. So it is essential to measure and monitor the healthcare processes enabled by the EMR system and provide feedback on the need to reengineer the healthcare system.

Organization of Monograph

In Chapter Two, we discuss our findings from a series of interviews with MHS senior leadership regarding their perceptions of the strategic value of AHLTA. The interviews covered both the appropriate dimensions of value—the strategic objectives—and how the leadership anticipated that AHLTA would affect those objectives. Chapter Three describes the analytic approach we recommend for estimating the strategic value of AHLTA. Chapter Four briefly reviews the available literature on the potential benefits of EMR systems. Chapter Five discusses detailed measures that can be calculated from the encounter-by-encounter data collected in AHLTA's clinical data repository. We focus on measures currently used to assess civilian healthcare quality and suggest approaches for developing new measures for those strategic objectives not well covered in civilian measure sets. Chapter Six discusses how to produce high-level measures by aggregating, summarizing, and selecting detailed measures. High-level measures convey "digestible" information to the senior leadership regarding MHS performance. Chapter Seven concludes the monograph with some comments about implementing and using our strategic measurement framework. A catalog of existing quality measures and their relevance to both AHLTA and to MHS strategic objectives is included on RAND's external Web site at the same URL as this monograph. The appendix describes this catalog.

CHAPTER TWO

MHS Senior Leaders' Views on the Value of AHLTA

Senior leaders of the MHS will be the primary customers for measures of the performance of the MHS and the strategic value of AHLTA. In the first phase of the project, therefore, we conducted a series of 14 semi-structured interviews with senior MHS leadership regarding their perceptions of the strategic value of AHLTA. Participants included senior leaders in OASD Health Affairs, the Health Affairs Office of Transformation, and the surgeon generals of the three armed services or their representatives. Interviews were scheduled to last 45 minutes to one hour and followed a protocol approved by RAND's Human Subjects Protection Committee. We asked interview participants to discuss the following four topics in an open-ended fashion:

1. The participant's roles and responsibilities as they relate to AHLTA
2. The anticipated effects of AHLTA on strategic performance
3. The role of AHLTA in performance measurement
4. Their general perceptions of AHLTA.

In the following sections, we discuss the themes that emerged during our interviews.

Relevant Dimensions of MHS Performance

Participants generally accepted that the strategic objectives as depicted in the MHS Strategy Map (Figure 2.1) captured all the relevant dimensions of MHS performance. None of them suggested that any strategic objective was superfluous or that any important aspect of MHS performance was not captured by one or more of the strategic objectives. Based on this consensus, we anticipate that grounding measures of AHLTA's contributions to MHS performance in the current set of strategic objectives would meet with a high degree of acceptance. We found that participants who are senior leaders of OASD Health Affairs knew precisely which strategic objectives fall within their areas of responsibility. The assignments can be seen on the MHS Balanced Scorecard (OASD, 2007b).

Interview participants who are not within OASD Health Affairs do not have direct responsibility for achieving MHS strategic objectives. However, their organizations were all involved in developing the MHS strategic plan, and they told us that they supported it.

AHLTA's Potential

Participants considered AHLTA to be a clinical tool, fundamentally affecting clinical care processes one patient or one encounter at a time. They recognized that AHLTA could help clinicians improve the quality and efficiency of the care they delivered. At the same time, they anticipated that one could aggregate data from many encounters and see AHLTA's effects on several strategic objectives. We provide examples below.

- *Individual medical readiness (IMR).* Participants cited the eventual role of AHLTA in improving the IMR rate through point-of-care clinical reminders and real time access to medical records. AHLTA is not currently designed to allow patients access to their health records, but if such a capability were implemented, it would further facilitate IMR improvements.

Figure 2.1
MHS Strategy Map Showing the Strategic Objectives

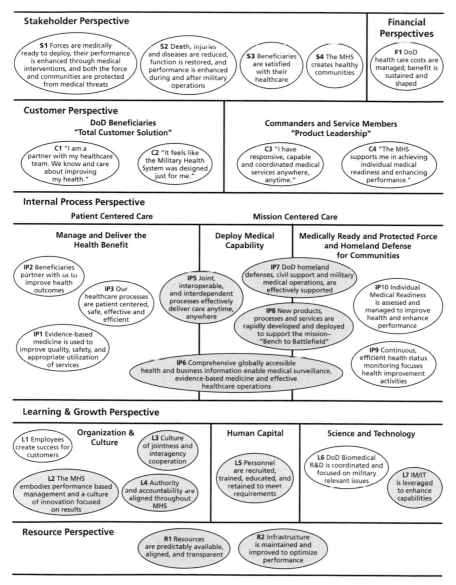

SOURCE: OASD (Health Affairs) (2007a).
RAND *MG680-2.1*

- *Measuring and promoting population health.* Participants were generally aware of AHLTA's future capacity to contribute to population health by serving as a surveillance tool, linking self-administered risk assessment data with clinical data, and hosting patient-accessible health records.
- *Quality and safety of clinical care processes.* Participants cited AHLTA's potential to transform clinical care processes through point-of-care alerts, reminders, and decision support tools.
- *Continuity of care.* Several mentioned AHLTA's ability to facilitate a team approach to patient care by making the entire medical record available to all team members. Several participants shared anecdotes about the value of centralized medical records in the aftermath of Hurricane Katrina.
- *Public health surveillance.* AHLTA could support disease and public health surveillance with more complete clinical data than are available from current sources. AHLTA requires doctors to enter symptom information using structured terms, and this makes it possible to establish computable links between clinical data and information on health behaviors and risk factors.
- *Efficiency and cost saving.* Participants cited AHLTA's potential to produce cost savings through reduced duplication of testing and procedures. Several mentioned the ability to eliminate duplicate or "shadow" paper record systems that evolved over time to deal with the problem of missing medical records inherent in the MHS's highly mobile patient population. Also mentioned was AHLTA's contribution to quantifying the aspects of productivity that are unique to military medicine and the readiness mission of the MHS.

A few participants recognized AHLTA as a process improvement tool.[1] There are many schools of process improvement, but they all advocate some variation on the following: (1) Collect data on how the chosen process is being carried out and how it scores on the perfor-

---

[1]   Examples were cited during the Health Innovations Program (HIP) awards session at the 2007 MHS annual conference (MHS Conference, 2007f).

mance measure; (2) formulate an approach to improving the process; (3) test the approach, adjusting and refining it as needed; and (4) implement the approach throughout the organization.

Several participants also suggested that the military leadership was hoping for too much from AHLTA. Some thought that AHLTA generates more information at a faster pace than the MHS is currently equipped to process and act upon. Also mentioned was the fact that AHLTA will be used only by providers in the direct care system, and records generated by providers in the purchased care system will not generally be available. Nor will providers of purchased care have access to the AHLTA records of their DoD patients.[2] Others noted that the MHS faces challenges that are not amenable to AHLTA-related "fixes." Examples included the short tenure of MTF commanders and the complex organization of the MHS, which comprises three healthcare systems, one for each military service.[3]

Present Status of AHLTA

While all participants felt that AHLTA would ultimately lead to more efficient, higher-quality healthcare, they saw AHLTA as too immature at present to have achieved much of its potential. But they also saw progress being made.

All participants recounted specific examples of their organization's experiences with implementing AHLTA, grappling with limited functionality, and managing AHLTA's steep learning curve. We commonly heard that AHLTA was initially resisted and then begrudgingly became standard operating procedure. Participants also noted that providers had begun to reach a level of comfort with the system and had begun to experience tangible benefits from it. Ground-level complaints

---

[2] The interviewees did not attempt to describe the size or consequences of this phenomenon, but see our discussion in Chapter One.

[3] Again, the interviewees did not discuss the reasons that these factors might limit the benefits of AHLTA, but we can speculate that they might undermine the exercise of the strong leadership that the quality improvement literature so often cites as an important success factor.

about AHLTA have evolved from initial resentment about having to use the system at all to unmet demands for enhanced functionality and speed.

We heard from several participants that AHLTA's value was presently limited in some ways by the fact that it was poorly linked to care delivered in a theater of operations.[4] Other participants pointed out that the MHS purchased a substantial amount of care from the private sector, and no good means existed for AHLTA users to obtain information about the purchased care a patient had received.[5]

Strategic Value of AHLTA

Only a few participants mentioned AHLTA's potential strategic value—that AHLTA might help high-level managers do their jobs better. These few participants had either immediate operational need for AHLTA capabilities that are currently absent or underdeveloped or were charged with the development and implementation of the performance measurement systems. They suggested that AHLTA could serve as a source of more and better data for measuring the strategic objectives identified in the MHS Strategic Plan and that this would give high-level managers enhanced opportunities to align incentives and accountability. See Table 2.1 for specific examples cited during interviews of the technical capabilities and leadership tasks required to optimize AHLTA's strategic value.

Participants with a strategic perspective also recognized that optimizing AHLTA's contribution to MHS performance required both technical and cultural change at all levels of the MHS. Table

---

[4] AHLTA access from the theater has evolved rapidly since the time our interviews took place.

[5] Discussions with the Clinical Information Technology Program Office (CITPO) staff suggest that the technology required for purchased care providers to enter clinical data into the CDR exists. However, there is currently no incentive for them to do so, particularly when EMR systems are still rare among commercial healthcare providers. The MHS also participates in the National Health Information Network, a federal initiative to foster the adoption of uniform standards for health information technology across private and public sectors.

2.2 provides specific examples of required changes cited by interview participants.

Table 2.1
Examples of Cited Technical and Leadership Contributions Required for AHLTA to Improve Strategic Performance of the MHS

AHLTA's Contributions	AHLTA Technical Capability	MHS Leadership Tasks
Information for decisionmakers	Structured text yields detailed, computable data through CDW/CDM Linkable to other data sources, e.g., cost, DEERS[a]	Establish a measurement framework Establish reporting and monitoring capacity Establish analytic capacity
Process improvement methodology	Comprehensive, accessible patient information Point-of care clinical support Real-time computation of visit value	Ensure system accessibility and stability Ensure AHLTA functionality Conduct user training
Alignment of incentives and accountability	Adaptable parameter settings, e.g., visit value Modifiable components and functions	Establish policies and procedures to improve performance

[a] DEERS = Defense Enrollment Eligibility Reporting System.

Table 2.2
Examples of Cited Technical and Cultural Changes Required to Optimize AHLTA's Contribution to MHS Performance by Organizational Levels

Type of Change	Top Level	Medium Level	Clinician Level
Technical	Establish and maintain analytic capacity	Ensure 100% system reliability and adequate bandwidth Create interoperability	Drag and click to produce structured text Personalize "tricks and shortcuts"
Cultural	Institutionalize use of information to guide decision-making Implement policies and procedures to optimize system potential	Improve appointment scheduling and patient flow Focus on prevention, wellness Embrace new patient roles and responsibilities	Accept computers in the exam room Assume burdens to improve system function Accept reduced professional autonomy

Conclusions

We take two lessons from these interviews. First, we should measure the value of AHLTA in terms of its effects on the MHS's strategic objectives (Figure 2.1). The leadership perceived that improvements in individual medical readiness, beneficiary satisfaction, the quality of care, and the health of communities are important potential effects of AHLTA, even if they do not save money.

Second, the leadership recognized that the current capabilities and user acceptance of AHLTA does not reflect AHLTA's permanent state. They expected AHLTA to generate more value in more different ways in the future than it can at present. This expectation implies that the methods for measuring AHLTA's value will need to evolve just as AHLTA itself does.

CHAPTER THREE

Analyzing the Effect of AHLTA on the Strategic Performance of the Military Health System

Federal law requires that DoD measure the value of information technology in terms of its contribution to the core mission of the adopting agency. To comply, the MHS must measure AHLTA's value in terms of its effect on strategic outcomes relating to efficiency, safety, readiness, and health. In this context, AHLTA's effect is the difference between outcomes achieved in an environment that includes AHLTA versus outcomes achieved in an environment that does not include AHLTA.

Measuring AHLTA's effect in this way presents two main challenges. The first is the formulation of a strategy for documenting MHS performance in the absence of AHLTA in a manner that is directly comparable to MHS performance in the presence of AHLTA. However, as we discuss below, simple comparisons of MHS performance prior to and subsequent to the implementation of AHLTA can be a misleading measure of AHLTA's effect because AHLTA is only one of many changing factors influencing the MHS during the time of its implementation. The second challenge is the development and implementation of relevant outcome measures, which we discuss in Chapters Five and Six.

The treatment-outcomes[1] model is the standard methodology for deriving quantitative measures of program effects in the face of the two

---

[1] Throughout this report we use the word *treatment* as it is used in the program evaluation literature: It refers to the action or intervention whose effects on outcomes is to be estimated. This is more general than the use made of the word in healthcare, where a treatment is something done to a patient to affect his health status.

challenges above (Mohr, 1988; Rossi, Freeman, and Lipsey, 1999). The treatment-outcomes methodology has four key elements:

1. *Outcome measures* capturing valued domains of system performance plausibly influenced by the presence of AHLTA. As described in Chapter Two, we determined from the interviews that these should be measures of the MHS strategic objectives (see Figure 2.1).
2. A *treatment* describing a way of using AHLTA that is expected to influence one or more outcome measures.
3. A *logic model* describing, in qualitative terms, the mechanisms or processes through which outcome measures are influenced. The logic model must describe not only the mechanisms through which uses of AHLTA (i.e., treatments) operate but also any important confounding influences.
4. An *evaluation design,* which is the strategy for separating a treatment's quantitative effect on outcome measures from the confounding effects of other factors.

As we describe in the sections below, implementing a treatment-outcomes model to obtain quantitative measures of AHLTA's effect is very data intensive. It may not be practical to measure AHLTA's effects quantitatively in many circumstances. As an alternative, we also discuss the use of logic models as a framework for gathering indirect evidence and substantiating the existence of a cause-and-effect relationship between a program and high-level, strategic outcomes (GAO, 1998; Greenfield, Williams, and Eiseman, 2006).

Outcome Measures

Outcome measures, as the term is used in Mohr (1988) and Rossi, Freeman, and Lipsey (1999), capture valued domains of system performance. In the current context, the valued outcome domains are the strategic objectives described in the MHS Strategic Plan (OASD, 2007a). The objectives as listed in the strategic plan, however, are too general and abstract to be used directly in measuring AHLTA's effect on performance. The MHS Balanced Score Card (OASD, 2007b) specifies quantifiable outcome measures for some of the strategic objectives. As an example, consider the MHS strategic objective S1: "Forces are medically ready to deploy . . . ." This could be measured by the fraction of active duty service members who meet all the criteria for IMR. We discuss measures to quantify strategic objectives in Chapters Five and Six.

Treatments

To make estimating AHLTA's value tractable, we have recommended breaking the problem into many small pieces, one for assessing each specific way of using one or more of AHLTA's features and functions under defined circumstances to influence valued outcomes. Each specific use of AHLTA will thus be a treatment to be evaluated using the treatment-outcomes framework. Continuing the example from the previous section, a treatment designed to influence the IMR rate of active duty service members could be a system of alerts and reminders that make clinicians who treat active duty service members aware, at the point of care, of the specific actions that must be taken to meet medical readiness requirements.

In Chapter Four, we discuss findings from the published literature on the various ways that EMR systems have been found to improve health system outcomes. These findings will highlight the features and functions of AHLTA upon which formal evaluations of AHLTA's effects would likely focus. This list will require updating as AHLTA continues to receive upgrades and enhancements and clinicians continue to learn new ways to use AHLTA and the information it generates.

Logic Models

Logic models trace the cause-and-effect relationships through which treatments influence outcomes of interest through a series of intermediary steps or sub-outcomes. Explicitly identifying the cause-and-effect relationships in this way is important for making the case that observed relationships between AHLTA-related features and functions and outcomes are causal and do not arise by chance. Logic models also play an important role in helping analysts understand the data required to compute AHLTA's effects quantitatively.

Carrying the IMR example further, a logic model would trace the key steps in the process through which a reminder to a clinician helps to speed the medical readiness process. Key elements might include the following:

- Technical capabilities: The reminder system is implemented.
- Treatment: Clinician receives reminder during sick visit.
- Sub-outcome 1: Clinician (re)notifies patient of his/her readiness status.
- Sub-outcome 2: Patient schedules required services.
- Sub-outcome 3: Patient receives required services.
- Outcome of interest: Patient is declared medically ready.

In specifying a logic model based on this example, it would be important to include pathways through which other factors (activities or initiatives occurring at the same time as the readiness alert features of AHLTA are being implemented) could influence the sub-outcomes and/or the outcome of interest. Failure to allow for the influence of other factors at this stage will preclude the consideration of those factors as possible confounding influences during the analysis that follows.

Each treatment requires its own logic model. However, we expect that many treatments will have very similar logic models. For example, any treatment that involves alerting or reminding clinicians about anything will look much the same as the example above.

Evaluation Design

A logic model provides a qualitative description of how AHLTA-related treatments and other factors influence outcomes. It specifies the structure of the process but does not measure the size of the effect. The evaluation design defines a way to combine data with the structure of the logic model to estimate the quantitative effect of the treatment on the outcome in a way that controls for confounding factors.

In the context of the medical readiness example above, the result would be the difference between the IMR rates for patients for whom AHLTA is used to improve the IMR rate (the *treatment group*), versus the IMR rate for patients for whom AHLTA is not used in this way (the *control group*). Ideally, the treatment and control groups would be identical except for the use of AHLTA to improve the IMR rate. Because this ideal does not exist, evaluation designs strive to approximate it in one of three main ways, depending on the availability of data and the conditions under which the program is implemented (Mohr 1988; Rossi, Freeman, and Lipsey, 1999).

- **Pre-Post Comparison.** The simplest method is to compare relevant outcomes before and after the treatment is implemented at one or more sites. Here, the patients visiting clinics prior to the AHTLA-IMR intervention form the control group and the patients visiting the clinics after implementation form the treatment group. Under this design, it is not possible to know whether the observed change was the result of the treatment or of other factors. Accordingly, it is advisable whenever possible to use one of the other approaches.
- **Pre-Post Comparison with Control.** Comparing outcomes over time for sites in which the treatment has and has not been implemented helps to ensure that observed outcome differences are due to the program and not other factors. Pre-post comparisons at the sites implementing the AHLTA-IMR intervention (the treatment group) yield an estimate of the effects of the treatment plus other factors; pre-post comparison at the sites without the treatment (the control group) yield an estimate of the effects of other factors

only. However, this method does not fully rule out differences between the treatment and comparison sites other than the presence of the treatment being evaluated, as the source of observed differences.

- **Pre-Post Comparison with Randomization.** Randomization of patients to sites implementing a treatment helps to ensure that treatment and comparison groups are similar. Thus, randomization can strengthen confidence that observed outcome differences are due to the program being evaluated and not other factors. This approach is considered the strongest of the three and should be adopted when possible.

Computation of AHLTA's Effects

Once measures of the treatment, outcomes, and confounding factors suggested by the logic model are in hand, analysts can use regression models to compute quantitative measures of AHLTA's effects. In essence, regression models yield estimates of the difference in the mean outcome for the treatment group and the control group, controlling for the effects of other factors. The complexity of the regression model required to obtain a valid estimate of AHLTA's effects depends on the extent to which confounding factors are controlled directly by the evaluation design. In general, more-rigorous designs that form treatment and control groups through randomization will be less complex than those that rely on pre-post differences alone to isolate treatment effects. The complexity of the regression models needed to estimate treatment effects under less-rigorous designs stems from the need to control, or adjust, the comparison of means for confounding factors.

Challenges to Measuring AHLTA's Effects on MHS Performance

Even with a set of well-specified outcome measures, the analytic strategies above can be challenging to implement, and the results these strategies yield can be difficult to interpret. In this section, we describe a number of confounding factors we identified during the course of our study that complicate the quantitative measurement of AHLTA's effects. Ignoring these factors in measuring AHLTA's effects could result in misleading conclusions.

Unstable Implementation Environment

A number of factors were influencing the MHS during AHLTA's implementation. These factors include demands for healthcare driven by the conflicts in Iraq and Afghanistan. AHLTA was not implemented in a manner (e.g., a sequenced rollout of AHTLA features and functions, randomized to outpatient clinics) that facilitates the isolation of AHLTA-relevant treatments from other factors influencing MHS performance.

Uncertain Technical and Organizational Adherence

To have its intended effect on system performance, AHLTA must be implemented and maintained properly, equipped with required technical capability, and used as intended. Investigators measuring AHLTA's effects must also know precisely when the AHLTA-related treatment has been "turned on" and whether clinicians and patients implement and adhere to it as intended.

Lack of Baseline Measures for Intermediate Outcomes

AHLTA has changed, and will continue to change, both in the processes by which clinical care is delivered and in the way patient encounters are documented. For example, procedure codes were assigned manually before AHLTA was implemented, but AHLTA calculates them automatically based on the structured terms clinicians use when they document a patient encounter. (This is "encounter coding support," one of AHLTA's Block 1 functions listed in Chapter One.) It is expected that identical encounters will sometimes be coded differently pre-AHLTA than post-AHLTA. This makes it difficult to determine how much of a pre-post difference is due to a change in actual practice as opposed to a change in documentation.

Lack of External Benchmarks

The Clinger-Cohen Act mandates that the performance of information technology investments be assessed against benchmarks established by comparable private and public sector organizations. However, opportunities for doing so are limited. Relatively few private health systems have implemented EMR. Large health systems that have implemented EMR, such as the Veterans Health Administration (VHA) or Kaiser Permanente, do not necessarily have comparable caseloads, organizational missions, and EMR functionality.

Partial Reliance on Direct Care

The fact that AHLTA is implemented only in the direct care systems and that MTFs provide only a fraction of the medical care provided to MHS beneficiaries influences the measurement of AHLTA's contribution to MHS performance in two ways. First, use of purchased care means that less utilization is subject to the influence of AHLTA. Thus, partial reliance reduces the potential magnitude of benefits attributable to AHLTA. Second, partial reliance can influence the validity of performance measures. For example, an AHLTA-generated reminder to clinicians to encourage patients to undergo preventive care screening may generate the intended response, but the screening may take place in the purchased care system. In this circumstance, a performance indicator that misses the resulting purchased care utilization would understate AHLTA's influence.

Threats to the validity of performance measures generated by partial reliance are in no way unique to the MHS. In fact, we argue that such threats are less serious in the context of AHLTA because the MHS administrative data systems record utilization that occurs outside of MTFs for the large subpopulation of TRICARE Prime enrollees. By contrast, the VHA has no way to track the use of non-VHA care by veterans who are not eligible for Medicare. Likewise, hospitals and medical groups cannot measure care that is not directly provided by them.

The presence of these factors will degrade, though unevenly, the quality of estimates of AHLTA's effects on MHS performance. That is, it may be impossible to obtain valid and statistically reliable estimates of some effects, while satisfactory estimates of other effects may be well within reach. This can only be determined case by case. The situation could be improved by collecting data describing the details of AHLTA's implementation, technical capabilities, and user compliance. As explained previously, sophisticated analytic and statistical approaches can mitigate the confounding effects of some factors.

Event Chains Based on Logic Models

The process used to measure program effects quantitatively is very data intensive. We imagine that in many cases suitable data may not be available. In such cases, logic models can be used as a framework for marshalling a "chain of evidence," both quantitative and qualitative, to substantiate the existence of a cause-and-effect relationship between a program and high-level, strategic outcomes. The Government Accountability Office (GAO, 1998) outlined such a recommended approach to specifying event chains in a published guide aimed at helping senior managers to develop strategies for measuring the performance of information technology investments.

Although event chains do not yield quantitative estimates of program effects, they rely heavily on quantitative data, trends in outcome data in particular. Thus, the performance measures we discuss in subsequent chapters can be employed, regardless of whether data to support quantitative measures of AHTLA's effects are available. In the context of the IMR example we have been using, the specification of an evidence chain based on the logic model described above might contain the following elements:

- Statistics describing the state of AHLTA implementation
- Documentation of training sessions aimed at teaching clinicians to use AHLTA properly
- AHLTA-generated IMR alerts and reminders that includes information about the training content, number of clinicians trained, geographic location of training sessions, and clinician satisfaction with training

- AHLTA-generated counts of the number of instances in which clinicians received reminders during sick visits
- Data on scheduling and receipt of IMR-related service
- Trends in IMR rates by MTF location and AHTLA implementation status.

We expect the event chain approach to prove valuable as MHS's capacity to analyze AHLTA data grows and as the commercial sector gains more experience with EMR technology and with measuring its benefits. However, as experience analyzing EMR data grows within the MHS and externally, stakeholders may demand quantitative assessments of EMR investments.

Summary

We have described approaches for measuring AHLTA's contributions to MHS performance that are grounded in a treatment-outcomes framework. We do not recommend applying these approaches to measure AHLTA's aggregate effect on the MHS as a whole or on any specific measure of MHS performance. Instead, we recommend measuring the effect of specific features and functions and functions of AHLTA under defined circumstances. We recommend this approach because AHLTA is capable of being used in an enormous number of ways to influence a wide variety of MHS activities and because AHLTA's influence will depend on how it is implemented and used.

We consider each specific use of AHLTA as a *treatment* and any valued domain of MHS performance as an *outcome*. The first step in assessing AHLTA's effects is to posit a logic model that qualitatively explains the mechanism or cause-and-effect chain by which the treatment influences the outcome. Subsequent steps involve fitting the logic model to data in order to quantitatively estimate the size of the influence.

Implementing this framework to obtain quantitative estimates of AHLTA's effect in the current environment will be challenging. Measures of MHS performance are likely to be influenced by many factors in addition to uses of AHLTA. Many of these other factors will be uncontrolled and even unmeasured. It will require considerable analytic sophistication to isolate the contributions of AHLTA in such a "noisy" environment. As the MHS's capacity to analyze AHLTA-generated data grows, it may be more practical (though potentially less convincing) to use logic models to specify evidence chains to demonstrate AHTLA's effects. Either approach will require the MHS to measure the strategically relevant outcome measures that we identify in Chapter Five.

To apply the treatment-outcomes methodology, it is necessary to define treatments and specify outcome measures. The next chapter discusses the kinds of treatments—the specific uses of AHLTA under defined circumstances—that the published literature suggests will be worthwhile to consider.

CHAPTER FOUR

Potential Benefits of EMR Systems

In this chapter, we discuss how lessons from the literature on EMRs can apply to the assessment of AHLTA. We briefly describe what is known from the literature about the specific uses of EMRs systems that produce benefits and the mechanisms by which those benefits arise. The specific uses provide us with an initial list of AHLTA-related treatments (as defined in Chapter Three) whose benefits we can estimate. The mechanisms provide us with a basis for constructing a logic model for each treatment.

An EMR is an information system. It provides a platform for collecting, storing, manipulating, and communicating information. There are some savings simply from doing these functions more cheaply than can be done with paper records. But the major benefits stem from the fact that clinicians get more complete information more rapidly than with paper records. If they make good use of the information, then healthcare can be made better on all the dimensions identified by the Institute of Medicine (IOM, 2001a); it can become more effective, more efficient, more equitable, more patient-centered, safer, and timelier. In other words, to take advantage of the potential benefits of the EMR system, clinicians must change the way they do their work. The EMR system enables or facilitates those changes, but it does not guarantee that change will actually happen or how quickly it will occur.

Many people treated in the civilian healthcare system see multiple providers, but those providers only rarely share their patients' medical records with one another. Fragmentation of the medical record is most prevalent when those records are kept on paper, but it remains a problem even for providers with electronic records. The literature we cite in this chapter does not specify the completeness of each EMR. It is reasonable to expect, however, that not all data used in each study came from patients whose records were complete. Nevertheless, these civilian EMRs have value and their value can be estimated. As we discussed in Chapter Three, an EMR has the most value for patients for whom the provider and health system have a reasonably complete record. This can happen, for example, if the patient happens to see only providers in a single group practice.

As we discussed earlier, AHLTA will not provide complete medical information about all MHS beneficiaries, even after the last planned features and functions (Block 4) have been deployed. Beneficiaries who receive some care from the purchased care system or from out of the system entirely will have less complete records in AHLTA than beneficiaries who receive all their care from the direct care system. But even now, AHLTA makes more information available about many MHS beneficiaries than was available before its deployment. The literature about civilian EMRs suggests, therefore, that AHLTA can even now be providing measurable value.

Effects on Efficiency

Girosi, Meili, and Scoville (2005) reviewed the effects an EMR system can have on efficiency. Some tasks can simply be done with less time and effort. Hospitals in the United States could save $15.2 billion per year by reducing unproductive time of nurses and eliminating the handling of paper records. They estimate that physician groups could save $3.6 billion per year in reduced transcription costs and chart pulls. An EMR system can also reduce duplication of laboratory tests and diagnostic images (e.g., X-rays, MRIs). They also estimate national outpatient savings of $5.8 billion per year and inpatient savings of $3 billion. Participants in some of our interviews said that duplication rates in the MHS might be higher than the national average, because military patients move from one provider to another more often than civilians. The EMR system can also advise the prescribing physician which drugs are covered by the patient's formulary. Girosi, Meili, and Scoville estimate the national savings from adhering to formulary guidelines at $12.9 billion for outpatients and $3.7 billion for inpatients. The system cannot compel the physician to prescribe the cheaper alternative, but it can require the physician to acknowledge that he has been advised that the cheaper alternative exists and has chosen the more costly drug anyway.

Physicians need to have the medical record in front of them when they call a patient to discuss the results of a test. The EMR system makes the medical record much more available to the physician than a paper record, and hence reduces callbacks. It should be noted that providers do not always save time, especially in the first few months after an EMR system is installed (Miller and Sim, 2004). Indeed, a project to install an EMR system can fail if it slows physicians down (Ornstein, 2003). Thus, one should expect a publication bias in favor of EMRs that deliver physician time savings.

Potential MHS-specific effects on efficiency are the following:

- Streamlining the medical separation and retirement processes by reducing time spent assembling paper records from diverse locations in preparation for review
- Streamlining the medical readiness process because clinicians are able to assess readiness at the point of care and active duty personnel can access their personal health records.

Effects on Individual Provider Performance

If used as intended, reminders and alerts delivered to clinicians can improve their compliance with guidelines (Asch et al., 2004; Chaudhry et al., 2006; Jha et al., 2003). McGlynn et al. (2003) developed 439 indicators of healthcare quality, and by medical record abstraction (a costly process) determined that the indicated service was provided on just over half the occasions it should have been. Each indicator can serve as the basis for a reminder. For example, a clinician could be reminded that the diabetic patient he is seeing today is due for an

HbA1c test or a retinal examination. An EMR can be configured to generate these reminders automatically if it specifically collects information required to determine whether a given reminder is indicated. Due to its emphasis on structured text, AHLTA should be capable of supporting an exceptional range of reminders, alerts, and guidelines.

Similarly, an EMR function called computerized physician order entry (CPOE) can improve the quality and safety of care that is delivered. Improvements in the outpatient setting can occur through ePrescribing, which is one form of CPOE for ambulatory care. The improvement happens because the EMR/CPOE system can check for (a) other drugs the patient is taking, (b) allergies, and (c) indications of impaired renal or liver function. Similar improvements occur in the inpatient setting (Chaudhry et al., 2006).

Care Coordination Among Providers

An EMR that gives multiple providers access to the same patient's record can help coordinate that patient's care. The EMR can make each provider aware of the others' decisions—such as what drugs the other providers have prescribed or what tests they have ordered. Generally, an EMR is accessible only to providers within the same organization, and AHLTA is no exception. But in the case of AHLTA, the organization is much larger and more comprehensive than any other.

The MHS should find that AHLTA helps maintain the continuity of care for military personnel and their families when they are reassigned and when military personnel return from a theater of operations with an injury or disease. We were told that many MTFs maintain, or have maintained, duplicate or "shadow" record systems in order to overcome problems associated with missing or incomplete paper medical records arising from the frequent relocation of highly mobile MHS beneficiaries. Unfortunately, these duplicate records are not always complete or up-to-date. AHLTA should make this practice unnecessary.

Care Coordination Between Provider and Patient

Patients experience better outcomes if they participate constructively in their own care (Bigelow et al., 2005). For the chronically ill, this means monitoring their own symptoms (e.g., a patient with congestive heart failure would keep track of breathing difficulties, weight gain, and swollen ankles). For others, it means adhering to their prescribed medications, a healthy diet and exercise regime, and avoidance of unhealthy behavior (e.g., smoking, excessive drinking).

The Chronic Care Model (CCM) describes a process for improving primary care for chronically ill patients (Bodenheimer, Wagner, and Grumback, 2002). Implementing the CCM requires a clinical information system (an EMR) with the capability of providing decision support. In this context, the EMR can

- generate a registry of all patients with a particular chronic condition, which can be used to plan the care of individual patients and to conduct population-based care
- provide feedback to physicians on how each patient is performing on chronic illness measures (e.g., HbA1c for diabetics)
- provide reminders to help primary care teams comply with guidelines
- provide a mechanism to support a patient's self-care efforts, and to help the patient report results of home measurements and observations (e.g., blood glucose for diabetics or peak flow for asthmatics) when the EMR possesses a patient portal.

There is little direct evidence that patients whose physicians use an EMR are more successful at self-care than patients whose physicians use paper records. But it is widely accepted that the way to involve patients in their own care is through education and communication (Bodenheimer, Wagner, and Grumback, 2002; Haynes, McDonald, and Garg, 2002; Miller, 1997). It seems reasonable that an information system, such as the EMR system, could facilitate education and communication.

There are no plans at present to provide patients with access to any part of their AHLTA records, nor to enable AHLTA to send reminders to patients. However, the TRICARE Management Activity (TMA) offers a patient portal called TRICARE-Online, which allows patients to make appointments online and access health-related content. In principle, the portal could eventually incorporate two-way access to the AHLTA record—i.e., patients could see (parts of) their own medical records, and information in the medical record could be used to generate reminders to patients and to highlight or even deliver to them health-related content appropriate to their condition.

Process Improvement

As mentioned previously, to take advantage of the potential benefits of an EMR, clinicians must change the way they do their work. Process improvement refers to systematic methods for learning how work practices should change and then implementing those changes. There are a number of examples of EMRs being used in process improvement activities in the VHA and civilian health systems. In 1995, the VHA initiated an effort to systematically improve the quality of care it delivers (Demakis et al., 2000; Kizer, Demakis, and Feussner, 2000; Rubenstein et al., 2000). It makes heavy use of VHA databases and information systems, including data from the VistA EMR (Hynes et al., 2000). Evaluations show that the quality of care delivered by the VHA has indeed improved (Asch et al., 2004; Jha et al., 2003).

Kilo (2005) describes a medical practice, Greenfield Health in Portland, Oregon, that "serves as a research and development laboratory for ambulatory system design." The paper starts with the bold statement: "Fifty percent of ambulatory care visits are unnecessary," and goes on to describe how the practice replaces a large fraction of in-person patient visits with secure electronic communication.

As mentioned in Chapter One, the MHS is establishing a clinical data mart and data warehouse. The CDM/CDW will contain the information from AHLTA's clinical data repository, but organized in a way that provides the types of data required to implement process improvement initiatives. Process improvement can be seen as a four-step approach guided by the formulation of a logic model as described in Chapter Three. Using the IMR example from Chapter Three,[1] we outline the steps below in Table 4.1.

AHLTA can play two roles in supporting process improvement initiatives. First, it can generate the reminders based on information in the CDR about individual service members' IMR status and deliver those reminders to clinicians at the point of care. Second, data collected by AHLTA can be used to support process improvement steps 1–3. If the IMR rate in fact improves in conjunction with the implementation of this initiative, it is reasonable to conclude that AHLTA merits some of the credit.

The roles of individual clinicians, high-level managers, and AHLTA can vary widely depending on the management style of the leadership. AHLTA's role may be relatively minor and unobtrusive in a situation in which managers use AHLTA/CDR-generated data to report to individual clinicians how their performance measures up against their fellow clinicians, or against a performance benchmark, and leave it to the clinician to decide what to do about it (steps 3 and 4). In the example above, if a clinician learns in a staff meeting that

Table 4.1
Process Improvement Steps and Examples

Process Improvement Step	IMR Example
1. Gather data.	Obtain information on IMR rates by clinician or the characteristics of active duty personnel.
2. Formulate an approach to improving the process using the information gathered in the first step.	Send reminders to clinicians.
3. Pilot-test and refine the approach.	Experiment with wording and linkages to appointment scheduling system.
4. Implement the approach throughout the organization.	Send reminders to all qualifying clinicians at finalized intervals.

[1] The example addressed the readiness of active duty service members to deploy. Guard and Reserve members are also subject to IMR criteria, but they are only eligible for MHS benefits if they are called to active duty for more than 30 days.

last quarter his patients were less medically ready than his colleagues' patients, the next quarter he may devote greater attention to readiness than in the past. AHLTA can also be used to create a tighter feedback loop relating clinical practice more closely to the receipt of performance information through its ability (1) to generate and deliver targeted reminders in real time and (2) to update clinicians in real time regarding their progress toward a benchmark. In a more extreme case, AHLTA could "grade" a clinician on the basis of readiness-related activities during a particular encounter and factor the grade into a clinician's performance record that could be viewed by the physician at the end of an individual encounter.

Public Health Surveillance

The MHS currently has a disease and public health surveillance system called Electronic Surveillance System for the Early Notification of Community-based Epidemics (ESSENCE) (MHS Conference, 2007b). AHLTA could support ESSENCE with more-complete clinical data than are available from current sources. This is because AHLTA requires doctors to enter symptom information as structured text and permits computable links between clinical data and information on health behaviors and risk factors.

Summary

The benefits of EMR systems have been reported in the literature. These reported benefits suggest the kinds of AHLTA uses that could serve as a starter set of treatments to evaluate using the methods described in Chapter Three.

There are minor benefits from automating activities that were previously manual, such as storing and retrieving medical records. But most benefits occur when clinicians change the way they do their work.

When EMRs generate reminders and warnings at the point of care, clinicians comply better with standards and guidelines—assuming the reminders are appropriately integrated into the care process. Because several providers can access the same person's electronic record simultaneously, they can better coordinate that patient's care. By facilitating communication between provider and patient, an EMR can also help a patient become more active and effective in his own care.

Like any information system, the EMR can potentially support process improvement efforts. Process improvement refers to systematic methods for learning how to do something better and then implementing those methods. An information system contributes to process improvement by measuring aspects of the process (e.g., delays, resource consumption) and feeding the information back to the improvement team.

AHLTA may also produce a benefit through public health surveillance. Unlike other EMRs, AHLTA stores information about symptoms as structured terms, so AHLTA will be more capable in this role than other EMRs are.

As mentioned at the close of Chapter Three, measuring the effect of AHLTA on MHS performance using a standard treatment-outcomes approach requires the specification of treatments and outcomes. This chapter discussed the kinds of treatments—the specific uses of AHLTA under defined circumstances—that the published literature suggests would be worthwhile to consider. The next chapter is the first of two that discuss outcome measures.

CHAPTER FIVE

Measures for the MHS Strategic Objectives

In this chapter, we discuss the outcome measures that are needed to measure AHLTA's effect on MHS performance, using a standard treatment-outcomes approach described in Chapter Three. Based on our conversations with MHS leadership, we concluded that outcome measures should quantify the strategic objectives described in the MHS Strategic Plan (OASD, 2007a) and displayed in the MHS Strategy Map (Figure 2.1). We pointed out, however, that the strategic objectives are too general and too abstract to be used directly in measuring AHLTA's effect on performance. For example, there are numerous aspects of the health of many individuals to be considered in assessing MHS strategic objective S4, "the MHS creates healthy communities." These include whether children treated in MTFs have received the recommended vaccinations, asthmatics and diabetics have their chronic conditions under good control, and MTF users are periodically screened for preventable diseases. In general, it takes a collection of many concrete and detailed measures to capture the full intent of a single strategic objective.

Below we identify concrete, detailed measures and associate them with strategic objectives. We will touch on all the strategic objectives, but we concentrate most attention on those that we believe AHLTA is most able to influence. To the extent possible, we identify measures included in well-accepted measure sets that are already in use in the U.S. civilian healthcare system, such as the Health Plan Employer Data and Information Set (HEDIS) measures defined by the National

Committee for Quality Assurance (NCQA)[1] or the Joint Commission's ORYX performance measurement systems.[2] These measures have been tested for both reliability (the degree to which the measure is free from random variation) and validity (the degree to which the measure is associated with what it purports to measure). Both the MHS and the individual services have borrowed heavily from these civilian measure sets in developing performance measures (OASD, 2007b; SGEGL, 2007; CMS, 2007).

For strategic objectives that are not well covered by existing measure sets, especially those strategic objectives that we think AHLTA can influence, we suggest possibilities for new measure development. We can suggest plausible measures in some of these cases, but they should be tested for reliability and validity before they are actually implemented.

Throughout this chapter, we refer to strategic objectives by their *indexes*, for example S4 or IP1. The strategic objectives are grouped into *perspectives*: stakeholder, financial, customer, internal process, learning and growth, and resources (See Figure 2.1).[3] Each strategic objective is indexed by an alphabetic prefix that denotes its perspective (S denotes stakeholder, IP denotes internal process, etc.) followed by a sequence number. The reader will find it helpful to keep a copy of the strategy map at hand while reading this chapter, in order to translate from the indexes to the text of the strategic objectives.

Detailed measures, such as the fraction of a defined population (e.g., individuals with diabetes type I and II) who do not meet a specific clinical standard (e.g., HbA1c level exceeds 9 percent), have limited utility for senior management. To achieve a comprehensive view of the performance of a healthcare system, one would need thousands

---

[1]   NCQA/HEDIS measures can be found at on the NCQA Web site. See NCQA (2007).

[2]   Joint Commission measures can be found on the commision's Web site (The Joint Commission, 2007).

[3]   The perspectives derive from the balanced scorecard (BSC) methodology that the MHS has used for strategic management since 2002. The original BSC methodology advocates providing views of an organization's performance from four perspectives: financial, customer, business process, and learning and growth. The MHS added two more, stakeholder and resource.

of such measures. To be practically useful, however, detailed measures must be carefully selected and/or summarized to give senior leadership a high-level view of strategic performance. The process of developing higher-level measures and summarization always loses information. It must be done in a way that conveys the most useful information while retaining the option to drill down into the detail to diagnose problems. We discuss strategies for building high-level measures from detailed measures in Chapter Six.

We have organized the remaining discussion in this chapter around the three themes described in the MHS Strategic Plan (OASD, 2007a):

1. Manage and deliver the benefit.
2. Create a deployable medical capability that can go anywhere, anytime with flexibility.
3. Provide a medically ready and protected force and medical protection for communities.

Theme I: Manage and Deliver the Benefit

This theme corresponds to the MHS task that is most like that of the civilian health system, that of providing quality healthcare to all DoD beneficiaries in a cost-efficient manner. The MHS Strategic Plan (OASD, 2007a) describes this theme as follows:

> **Manage and deliver a superb health benefit**—We build partnerships with our beneficiaries in an integrated health delivery system that encompasses military treatment facilities, private sector care and other federal health facilities including the Department of Veterans Affairs (VA). Globally accessible health and business information enables patient-centered evidence-based processes that are both effective and efficient (p.14).

The MHS Strategy Map identifies the strategic objectives associated with this theme (Table 5.1).[4]

AHLTA in its current state has the potential to influence each of these strategic objectives. The nature and magnitude of AHLTA's influence depends on the extent to which the organizational and cultural changes have occurred that are required to realize AHLTA's potential benefits (see Table 2.1). Simply implementing AHLTA's CDR has made beneficiaries' medical records available at all MTFs worldwide, and implementing AHLTA's CDW will make health information for analysis and population surveillance globally accessible (objective IP6). The benefits of EMR systems that we discussed in Chapter Four include many that fall within objectives IP1–IP3. We presume that if the process-related objectives IP1–IP3 improve, so will stakeholder- and customer-related objectives S3, S4, C1, and C2. AHLTA has been designed to generate diagnosis, procedure, and billing codes automatically from the documentation of a patient encounter. Codes from

Table 5.1
Strategic Objectives Associated with Theme I:
Manage and Deliver the Benefit

Index	Objective
S3	Beneficiaries are satisfied with their healthcare.
S4	The MHS creates healthy communities.
F1	DoD healthcare costs are managed; benefit is sustained and shaped.
C1	"I am a partner with my healthcare team. We know and care about improving my health."
C2	"It feels like the MHS was designed just for me."
IP1	Evidence-based medicine is used to improve quality, safety, and appropriate utilization of services.
IP2	Beneficiaries partner with us to improve health outcomes.
IP3	Our healthcare processes are patient centered, safe, effective, and efficient.
IP6	Comprehensive globally accessible health and business information enables medical surveillance, evidence-based medicine, and effective healthcare operations.

---

[4] OASD (2007a).

AHLTA should be more complete and more consistent than codes assigned by coding specialists working from free-text medical records, as was done in the past. Better coding should facilitate improved management of costs (objective F1).

Measures for Strategic Objectives Related to the Delivery of High-Quality Care (S3, S4, C1, C2, IP1, IP2, and IP3)

In this section, we discuss measures developed for civilian healthcare organizations. As we commented earlier, Theme I corresponds to the MHS task that is most like that of the civilian health system. Thus, these measures should be particularly relevant to the strategic objectives corresponding to this theme. However, none of the measures we have identified applies to financing of healthcare (objective F1) or to assessing the information infrastructure of healthcare organizations (objective IP6). This section, therefore, addresses the remaining measures from Table 5.1.

The Agency for Healthcare Research and Quality (AHRQ) sponsors the National Quality Measures Clearinghouse (NQMC), a database and Web site with information on specific evidence-based healthcare quality measures and measure sets.[5] As of April 11, 2007, the NQMC database contained 1,153 measures.[6] We have assembled a catalog of these measures in the form of an Excel spreadsheet, which appears on the RAND external Web site, at the same URL as this monograph.[7]

In essence, the NQMC database contains the universe of publicly available, ready-to-use healthcare quality measures. For example, it contains all the measures in the "starter set" of quality measures recommended by the IOM (2006, p. 204), the HEDIS measures, and the Joint Commission's measures. Each measure in the database must be accompanied by one or more of three kinds of evidence:

---

[5] NQMC (2007).

[6] This number overstates the number of unique measures. The same or very similar measures are often submitted by two or more organizations. For example, no fewer than four measures ask for the fraction of diabetic patients with hemoglobin A1c levels above 9 percent.

[7] http://www.rand.org/pubs/monographs/MG680/

1. a citation in a peer-reviewed article or report applying or evaluating the measure's properties; or
2. peer-reviewed documentation of the measure's reliability and validity; or
3. development, modification, or endorsement by an organization that promotes rigorous development and use of clinical performance measures.

According to the documentation on the NQMC Web site, 15 percent of the measures require only data from administrative sources (e.g., claims, eligibility), pharmacy data, or public health reports. These measures could probably be implemented for both direct and purchased care with data currently available to the MHS. Another 45 percent require, in addition, data elements typically available in laboratory reports, medical records, or patient registries. AHLTA should make it possible for the MHS to implement all or most of these measures for the direct care system,[8] but not for purchased care. The remaining 40 percent of the measures require special data collection, such as surveys. To the extent that the MHS is willing to field the necessary data collection efforts (it already conducts a number of surveys), these measures could be implemented for both direct and purchased care.

We associated each NQMC measures with strategic objectives (a full description of our method appears in the appendix). Each measure was associated with zero (168 measures), one (514 measures), or two (471 measures) objectives. Table 5.2 shows the number of NQMC assigned to each objective. All but a handful of the assignments were to the objectives considered in this section.

We next discuss the extent to which the NQMC measures provide good coverage of these strategic objectives. Objectives C1 and IP2 are both concerned with the degree to which patients participate effectively in their own healthcare. By themselves, the raw assignment numbers from Table 5.2 suggest that these objectives are not well covered

---

[8]   This should be true in general, but there are bound to be exceptions. Each measure requires specific data elements, and it may be that some measures require data elements that AHLTA happens not to collect.

Table 5.2
Counts of Measures by Strategic Objective

Index	Objective	No. of NQMC Measures
S3	Beneficiaries are satisfied with their healthcare.	108
S4	The MHS creates healthy communities.	213
C1	"I am a partner with my healthcare team. We know and care about improving my health."	21
C2	"It feels like the MHS was designed just for me."	88
IP1	Evidence-based medicine is used to improve quality, safety, and appropriate utilization of services.	482
IP2	Beneficiaries partner with us to improve health outcomes.	14
IP3	Our healthcare processes are patient centered, safe, effective, and efficient.	331
IP6	Comprehensive globally accessible health and business information enables medical surveillance, evidence-based medicine, and effective healthcare operations.	15
L5	Personnel are recruited, trained, educated, and retained to meet requirements.	16
	Total	1,456

by the NQMC measures. The MHS may wish to consider developing additional measures for these objectives. But unless and until clinicians use AHLTA to reach out to patients (e.g., via the TRICARE Patient Portal), it is hard to see how AHLTA can either influence these objectives or help to measure them.

Objectives S3 and C2 are concerned with customer satisfaction. All the NQMC measures that we assigned to these objectives are calculated based on survey data, sometimes augmented with administrative data. We see no means by which AHLTA could be used to gather satisfaction data, although proper use of AHLTA should indirectly influence these objectives. Accordingly, we gave low priority to identifying measures for these objectives.

Objectives S4, IP1, and IP3 are all concerned with how well healthcare providers do their jobs. Table 5.2 shows that provider performance is the primary focus of the NQMC measures. These are also the NQMC measures that AHLTA can influence most directly. In addition, it should be possible to calculate many of these measures automatically from the data in AHLTA's CDM/CDW. In fact, many of these measures, especially those assigned to objectives IP1 and IP3, could readily be turned into point-of-care reminders. The reminder says: "I see that condition X applies to your current patient. You should do Y." The measure calculates: "In what percentage of the instances when condition X applied did the provider do Y?"

Despite the large numbers in Table 5.2, the NQMC measures leave gaps and potential gaps in the coverage of objectives S4, IP1, and IP3. The IOM has identified gaps in currently available measures (IOM, 2006). Under the heading "Comprehensive Measurement" the IOM says:

> Current performance measure sets are far too limited in scope. The vast majority of current measures assess the quality of care in terms of effectiveness and safety. Only a few, limited measures examine timeliness and provide insight into patients' experiences, and hardly any adequately assess the efficiency or equity of care. Nor do measures adequately cover the entire human lifespan, as very few evaluate care for children, adolescents, or those at the end of life. Finally, too few measures exist that address matters particularly salient for the Medicare population, such as chronic obstructive pulmonary disease, stroke, dementia, and Alzheimer's disease (p. 89).

From the MHS perspective, some of these gaps may not be particularly important, and there may be gaps in coverage not mentioned here that loom larger for the MHS. We suggest that the MHS review historical claims data to find which conditions impact the military services most. New measures should be developed for any high-impact conditions that are neglected in the NQMC dataset. Under the heading "Longitudinal Measurement" the IOM (2006) says:

> The committee's emphasis on longitudinal measurement is based on two distinct concerns. First, both the U.S. fee-for-service system and the performance measures currently in use reinforce, although not intentionally, the separation of settings of care by design (i.e., ambulatory care, home health care, hospital care, and nursing home care). This emphasis on separate care settings

has several adverse effects, including fragmentation, lack of continuity, and poor communication. Second, the effectiveness of a care system should ideally be reflected in its capacity to prolong life, maintain or improve functioning and the quality of life, and achieve health outcomes with a high degree of patient centeredness and efficiency. Achievement of these results generally involves care that crosses boundaries, rather than actions of a particular caregiver at a specific point in time. Measurement that focuses only on such fragments of care misses too much of what really matters to patients. Rather, measure sets should concentrate on measures of continuity and transitional care, as well as on longitudinal assessments of health outcomes and costs (pp. 89–90).

One of our interviewees recommended that the MHS use "episodes of care" to incorporate care coordination (as opposed to fragmentation) into performance measurement. This individual described an *episode of care* as the total of all care provided for a specific medical problem or condition, i.e., all the care provided from the appearance of a medical problem until its resolution. For an acute condition, the resolution may be a cure. For a chronic condition, the resolution may consist of bringing the condition under control.

We agree that episodes of care, if one can identify them, are attractive units of care to analyze. Unlike a single patient encounter with a provider, an episode has a meaningful output (the resolution of the problem) that has a value. Meaningful quality, timeliness, and efficiency measures can be defined more readily for episodes than for single healthcare events. (Of course, there might still be reasons to examine individual events within the episode of care.)

It is difficult, however, to identify all the healthcare events that belong to an episode, particularly where not all care provided during an episode (i.e., purchased care) is recorded in AHLTA. Proprietary methods have been developed for constructing episodes from claims data,[9] but we know of no publicly documented and validated method.

---

[9] Episode Treatment Groups (ETGs) were developed by Ingenix; The Medical Episode Grouper (MEG) was developed by Thomson Medstat. See Ingenix (2007) and Thomson Healthcare (2007) for details.

Nonetheless, because the episode-based approach has the potential to yield highly clinically relevant information, it would be worthwhile for the MHS to investigate ways to construct episodes. Doing so—for even part of the workload—may be particularly informative.

Measures for Strategic Objective Related to Managing Cost (F1)

The strategic plan (OASD, 2007a) explains strategic objective F1 as follows:

> The MHS health care delivery system will be engineered to achieve optimal efficiency and mission effectiveness. The TRI-CARE benefit will reinforce appropriate use of resources and demand for services, and will engage the individual to actively manage his/her health (Appendix).

The civilian sector has little guidance to offer the MHS in measuring the cost or efficiency of medical care.[10] Although there are numerous peer-reviewed articles on healthcare efficiency, very little has been published on the reliability and validity of efficiency measures. The NQMC measures are quality measures, and none of them addresses cost or efficiency. Each organization—each purchaser, each hospital chain—tends to develop its own measures of efficiency, based on its own data sources, cost structure, and objectives.

MHS strategic planning staff have developed a high-level efficiency metric that measures total per member per month (PMPM) healthcare spending (MHS Conference, 2007d). The measure draws on standard methods for developing efficiency measures and illustrates many of the difficulties. The component measures that make up PMPM map to strategic objectives IP1, IP3, and F1. For IP1, the four measures are

- relative value units (RVUs) per beneficiary for ambulatory care
- relative weighted product (RWP) per beneficiary for inpatient care

---

[10] See for example, Academy Health (2006).

- prescriptions per beneficiary for pharmacy
- tests per beneficiary for ancillary services.

RVUs measure the resources or effort required to produce a given unit of service in terms of "physician's work input (e.g., mental effort, technical skill), the opportunity cost of specialty training, and the relative practice costs for each specialty" (Hsiao et al., 1988). An RWP is a DoD measure of workload that represents the relative resource consumption of a patient's hospitalization compared to that of other inpatients (ASD/HA, 2002).

These four resource intensity measures are not currently obtained from the CDR. Because AHLTA automatically assigns billing codes to patient encounters, the use of AHLTA to generate the four measures should result in RVU counts per beneficiary that are more consistent from one MTF to another. It will not, however, improve the consistency of RVU counts obtained from claims for purchased care. AHLTA Block 3 will be able to provide pharmacy and ancillary service utilization. AHLTA Block 4 will subsume the inpatient record and will report RWP counts.

The numbers of beneficiaries forming the denominators of these ratios are obtained from DEERS. Because different categories of beneficiaries can be expected to use different amounts of healthcare, the number of beneficiaries is adjusted for age, gender, and beneficiary category. It could be important to adjust these numbers to account for beneficiaries who receive part of their care from outside the MHS, if this is not done already. Failure to adjust for this factor could make PMPM values look smaller than they should.

There are four cost measures corresponding to each of these resource consumption measures; and they map to strategic objective IP3:

- cost per RVU for ambulatory care
- cost per RWP for inpatient care
- cost per prescription for pharmacy
- cost per test for ancillary services.

The cost measures are calculated as the ratio of an expenditure for a given service category to the number of RVUs, RWP, prescriptions, or tests (as appropriate) purchased by that expenditure. Expenditure data are obtained from the MHS's cost accounting system, Medical Expense and Performance Reporting System (MEPRS); the Pharmacy Detail Transaction Service (PDTS); and other sources. Separate cost measures are calculated for each MTF and for providers of purchased care, thus offering a way to compare the efficiencies of different sources of care.

However, accounting systems generally capture costs by expenditure category (e.g., salaries) and by cost center (e.g., clinical department). To estimate a cost per RVU (for example), the cost elements in the accounting system must be allocated to RVUs. It is easiest to allocate costs in proportion to RVUs, which amounts to assuming that every RVU produced by a particular cost center has the same cost. If this is not true, a change in the mix of RVUs delivered will change the cost per RVU.

The overall PMPM measure is obtained by multiplying each of the four resource intensity measures by its corresponding unit cost measure to derive a component PMPM, then summing the four component PMPM measures.

The MHS has plans to improve the PMPM measures in two ways. First, the MHS uses RVU codes borrowed from the civilian sector, and some services the MHS delivers have no civilian counterpart and thus are omitted from the RVU counts. This artificially inflates the cost per RVU factors. Dinneen (MHS Conference, 2006) identified a number of omitted categories, including assessment of individual medical readiness, medical boards, and military-unique training.

Second, the Defense Medical Human Resource System-Internet (DMHRSi) will soon be collecting more-accurate information on how clinical staff spend their time. Without this information, a clinician's salary cost is often attributed to his department of record. With it, salary costs can be allocated to the department in which the work was done.

The most fundamental problem with measuring cost and/or efficiency is that the MHS (in common with most enterprises) records costs at a more aggregate level—cost center and budget category—than that needed to properly answer many of the questions one wishes to ask. All the NQMC quality measures, for example, are ultimately calculated from data elements that apply to individual patients and can therefore be aggregated over arbitrarily defined groups of patients. This problem afflicts civilian healthcare (indeed, every sort of enterprise) as much as it does the MHS—and it will not be solved quickly.

Measures for Strategic Objective Relating to Global Access to Information (IP6)

The MHS balanced scorecard (OASD, 2007b) lists two measures for objective IP6: (1) percentage of all Standard Ambulatory Data Record (SADR) encounters that were completed in AHLTA and (2) satisfaction with AHLTA. The first is relevant to AHLTA because increases in the AHLTA-completed encounters increase the enterprise-wide availability of medical record data. User satisfaction is related to global access to information to the extent that more satisfied users record more AHLTA encounters and record them reliably. However, we reviewed no information that would either prove or disprove this proposition.

We suggest that these two measures be augmented with measures of information accessibility that AHLTA could influence. Such measures require lists of information that need to be accessible in order to perform health and public health–related work processes in a timely and complete fashion. For example, the IMR process requires information on active duty Army enlisted personnel not medically ready because of dental deficiencies. Information accessibility measures could be constructed on the model of NQMC measures developed by the British Medical Association that take the following form: "AHLTA can produce a registry of patients with chronic condition X." Such measures can be scored on a pass/fail basis.

Theme II: Create a Deployable Medical Capability

This theme corresponds to the ability of the MHS to physically treat deployed personnel outside the MTFs. The Strategic Plan (OASD, 2007a) describes this theme as follows:

> **Create a deployable medical capability that can go anywhere, anytime with flexibility, interoperability and agility.**—We monitor globally accessible health information and rapidly develop and deploy innovative medical services, products and superbly trained medical professionals. Our involvement in the full range of military operations includes assistance in civil support and homeland defense (p. 14).

Table 5.3 lists the strategic objectives associated with this theme in the MHS Strategy Map.

We anticipate that AHLTA can be used to influence all these objectives except IP8, which involves technology adoption. (We discussed objective IP6 earlier, under Theme I.)

Measures for Strategic Objectives Relating to Medical Care During Military Operations (S2, C3, and IP5)

As described in the Strategic Plan, objective IP5 has two parts. First, the MHS must be able to rapidly deploy medical capability to theaters of operation. Second, that medical capability should be joint, interoperable, and interdependent so that members of any service, anywhere in the theater or at home station, will receive the care they need. We believe the deployment of a theater version of AHLTA would improve the ability to deploy medical capability. Because AHLTA's interface is uniform across settings and because all services will use it, it should promote jointness and interoperability. We interpret interdependence to mean continuity of care, both within the theater and with the home station.

We suggest that, for the purposes of evaluating the contribution of AHLTA to MHS strategic performance, measures for objective IP5

Table 5.3
Strategic Objectives Associated with Theme II:
A Deployable Medical Capability and Homeland Defense for Communities

Index	Objective
S2	Deaths, injuries, and diseases are reduced, function is restored and performance is enhanced during and after military operations.
C3	"I have responsive, capable, and coordinated medical services anywhere, anytime."
IP5	Joint, interoperable, and interdependent processes effectively deliver care anytime, anywhere.
IP6	Comprehensive globally accessible health and business information enables medical surveillance, evidence-based medicine, and effective healthcare operations.
IP7	DoD homeland defenses, civil support and military medical operations are effectively supported.
IP8	New products, processes and services are rapidly developed and deployed to support the mission—"Bench to Battlefield."

should be measures of the continuity of care. Continuously available, coordinated, high-quality care should reduce deaths, injuries, and disease during operations and improve the care of service members who suffer wounds, injuries, or disease in-theater (objective S2). Simultaneously, commanders and service members should gain confidence that high-quality care will be available anywhere, anytime (objective C3).

One approach to assessing continuity of care would be to develop episode-level measures of treatment processes and health outcomes. We discussed the challenges presented by this approach above. As an alternative, we suggest the creation of a targeted list of AHLTA's technical capabilities that contribute to continuity of care in selected strategic domains (e.g., care of deployed forces, care of dependents with chronic conditions), development of measures demonstrating the deployment and active use of such capabilities. For example, consider a service member who needs medical care in a theater of operations. Measures relevant to this situation might include the following: (1) Does AHLTA have the ability to provide the first responder with the needed information about this service member (e.g., blood type, allergies, medications)? (2) Once the service member is transported to a field hospital, does the physician have the first responder's notes? (3) If the service member must be evacuated out of theater, does the destination MTF have information about previous treatment in-theater?

Measures of objective S2 can be patterned after the quality measures in the NQMC. The NQMC has essentially no measures for trauma care, whether physical or psychological. However, in addition to the NQMC, the National Guideline Clearinghouse (NGC) contains some guidelines related to treatment of traumatic injury. For example, the Brain Trauma Foundation submitted a guideline on the management of severe traumatic brain injury, and the Paralyzed Veterans of America submitted one on outcomes following traumatic spinal cord injury. While a guideline is not a measure of the quality of care, it may be possible to construct measures of the difference between a guideline and the care actually delivered.

We remind the reader that before a measure is adopted, it should be tested for reliability (the degree to which it suffers from random variation) and validity (the degree to which it measures what is intended). The suggestions we have just made have not been assessed in this way.

Measures for Strategic Objectives Relating to Homeland Defense and Civil Support (IP7)

Objective IP7 extends support for military operations (objective IP5) to assistance in civil support and homeland defense operations, such as disaster relief and management of pandemic influenza. AHLTA's role in facilitating this objective may be limited because it only maintains data on its own military beneficiaries treated in MTFs. To the extent that DoD beneficiaries are affected by such events, however, AHTLA will have a role in facilitating care and disease surveillance, as discussed earlier.

Theme III: Provide a Medically Ready and Protected Force

This theme corresponds to the ability of the MHS to assure the physical health of deployable and deployed forces. The Strategic Plan (OASD, 2007a) describes this theme as follows:

Provide a medically ready and protected force and medical protection for communities.—We continuously monitor health status, identify medical threats and find ways to provide protection and improve health for individuals, communities and the Nation. These surveillance activities focus our delivery of Individual Medical Readiness services to improve health and enhance human performance and make the environment safer so service members can withstand health threats in hostile settings. We develop countermeasures to protect against medical threats for the military force and the nation (p. 14).

Table 5.4 lists the strategic objectives associated with this theme in the MHS Strategy Map.

We anticipate that AHLTA can be used to influence progress toward all these objectives except IP8, which relates to technology adoption. (We discussed objective IP6 earlier, under Theme I.)

Measures for Strategic Objectives Relating to Monitoring Health Status (IP9)

The Strategic Plan (OASD, 2007a) explains objective IP9 as follows:

> Throughout the life cycle of a service member, the system will capture and analyze health information and identify health hazards associated with the military environment enabling focused health promotion and disease prevention (Appendix).

AHLTA can capture information on reported symptoms and diagnosed conditions as needed. To use AHLTA-generated information for health promotion and disease prevention, it must be linked with data on the larger at-risk population using data such as DEERS, surveillance databases, or registries of DoD beneficiaries with diagnosed health conditions. Using this type of linked data, AHLTA's contribution to health promotion and disease prevention can be measured using many of the measures in the NQMC set.

Table 5.4
Strategic Objectives Associated with Theme III:
A Medically Ready and Protected Force

Index	Objective
S1	Forces are medically ready to deploy, their performance is enhanced through medical interventions, and both the force and communities are protected from medical threats
C4	"The MHS supports me in achieving individual medical readiness and enhancing performance."
IP6	Comprehensive globally accessible health and business information enables medical surveillance, evidence-based medicine, and effective healthcare operations
IP8	New products, processes, and services are rapidly developed and deployed to support the mission—"Bench to Battlefield"
IP9	Continuous, efficient health status monitoring focuses health improvement activities
IP10	Individual medical readiness is assessed and managed to improve health and enhance performance

AHLTA can be used retrospectively to support epidemiologic investigations of the relationship between potential health hazards posed by military operations. To demonstrate AHLTA's contribution to such investigations, the MHS should maintain careful records documenting the timeliness with which AHLTA data is made available to investigators and the completeness and accuracy with which AHLTA records are linked to data documenting the exposures being investigated.

AHLTA can also play a prospective role in identifying the health hazards associated with military operations that have not yet been identified. The MHS already has a population surveillance tool, ESSENCE (MHS Conference, 2007b). We anticipate that incorporating data from AHLTA will improve the tool by making information on symptoms available in a more complete and rapid fashion compared with methods that rely on review of paper medical records and existing (non–clinically specific) administrative data.

Measures of AHLTA's contribution to surveillance should aim to document (1) improvements in surveillance capabilities resulting from AHLTA's implementation and (2) instances in which AHLTA implementation facilitated DoD's rapid response to a biothreat. Examples of such improvements might include expanded abilities to

- monitor a greater range of diseases and symptoms
- rapidly and accurately link symptom data to exposure data
- track exposed individuals over long periods of time and wide geographic areas
- alert clinicians at the point of care to document symptoms and conditions that are under active investigation.

Measures for Strategic Objectives Relating to Individual Medical Readiness (S1, C4, and IP10)

The military has precise criteria that define IMR status. To be medically ready, a service member requires a periodic health assessment, must not have a deployment-limiting condition, must need no or only nonurgent dental treatment, must have up-to-date immunizations, and must have the appropriate medical equipment (e.g., eyeglasses including corrective lens inserts for a gas mask, hearing aids). Each criterion relates to measures corresponding to strategic objectives IP10, S1, and C4.

IMR-related objectives can be influenced by features of AHLTA that provide the service member with access to his or her health information and send the service member and clinician reminders that a given action must be taken to become IMR-certified. However, we know of no current plans to enable AHLTA to send reminders to service members.

Because AHLTA is not yet linked to IMR-related data systems— e.g., the Medical Protection System (MEDPROS)—and does not yet host personal health records, we recommend that the MHS explore the feasibility of demonstrating the effect of AHLTA on the IMR process through a prospective study design of the type described in Chapter Three. Close coordination around the planned rollout of IMR-related capabilities could help maximize the validity and precision of such a study. The study could compare change over time in the following IMR-related outcomes for a defined population or cohort pre- and post- implementation of IMR-related AHLTA features:

1. Members of the relevant population who meet a given criterion
2. Members who we know do not meet the criterion
3. Members who we know do not meet the criterion and who have been reminded that their IMR status needs to be resolved
4. Members for whom it is not known whether they meet the criterion
5. Members for whom it is not known whether they meet the criterion and who have been reminded that their IMR status needs to be resolved.

The change over time in the ratio of Group 4 members to the population or cohort of interest could potentially measure AHLTA's contribution to the effective management of the IMR process (IP10). Change over time in the ratio of Group 1 to the whole defined population is a logical measure of changes in the overall level of readiness (S1). The change over time in the ratio (Group 3 + Group 5) / (Group 2 + Group 4) may be a reasonable proxy for the perceptions of active duty personnel regarding the supportiveness of the MHS in the IMR process (C4).

The Remaining Strategic Objectives

Tables 5.5 and 5.6 list the remaining MHS strategic objectives not discussed in the previous sections. Because AHLTA's contribution to these remaining objectives is indirect, we do not propose AHLTA-sensitive outcome measures for them. Nonetheless, we suggest that the MHS monitor and document these indirect relationships as a means of specifying logic models describing the process through which AHLTA influences MHS strategic outcomes. As we suggested in Chapter Three, well-specified logic models play a key role in the quantitative measurement of AHLTA's effects.

Table 5.5
Strategic Objectives for the Learning and Growth Perspective

Index	Objective
Organization and Culture	
L1	Employees create success for customers.
L2	The MHS embodies performance-based management and a culture of innovation focused on results.
L3	Culture of jointness and interagency cooperation.
L4	Authority and accountability are aligned throughout MHS.
Human Capital	
L5	Personnel are recruited, trained, educated, and retained to meet requirements.
Science and Technology	
L6	DoD biomedical R&D is coordinated and focused on militarily relevant issues.
L7	IM/IT is leveraged to enhance capabilities.

Table 5.6
Strategic Objectives for the Resources Perspective

Index	Objective
R1	Resources are predictably available, aligned, and transparent.
R2	Infrastructure is maintained and improved to optimize performance.

CHAPTER SIX

Strategic Management

In Chapter Five, we discussed outcome measures for the MHS strategic objectives. We pointed out that the strategic objectives are too general and abstract to be used directly in measuring AHLTA's effect on performance. Thus, it takes a collection of hundreds of concrete and detailed outcome measures to capture the full intent of a single strategic objective.

Senior leadership, however, need to know the degree to which the strategic objective is being achieved. They cannot afford to concern themselves with hundreds of details. In this chapter, we address the issue of summarizing the hundreds of detailed measures into a handful of strategic measures—i.e., aggregate indicators for a strategic objective. The IOM identifies the lack of strategic (in IOM terms "aggregate") measures as an important gap in the set of currently available quality measures (IOM, 2006, pp. 91–95). Our proposals for constructing strategic measures from detailed measures parallel theirs.

Inevitably, high-level, strategic measures (e.g., total PMPM cost) will contain less information than the original, detailed measures. For this reason, senior leadership should retain the option to drill down into the detail to diagnose problems. If one constructs the strategic measures as we describe, an audit trail will exist that makes drilling down possible.

As is true for the detailed measures from Chapter Five, we are primarily concerned with using the strategic measures as outcome measures for the treatment-outcomes methodology described in Chapter Three. Used in this way, they will be measures of the strategic value of various uses of AHLTA. But these measures play a more general role in MHS strategic management. The MHS strategic plan (OASD, 2007a) says:

> Senior leadership will conduct periodic reviews of MHS performance against our strategic targets and adjust activities and resources to continuously improve (p. 8).

The remainder of this chapter is divided into two sections. The first discusses approaches for constructing strategic measures from the detailed measures of Chapter Five. The second discusses how to use strategic measures for strategic management.

Constructing Strategic Measures from Detailed Measures

To build strategic or aggregate measures from detailed measures, it is important to understand the basic structure of a detailed measure. All the NQMC measures are built up from two items of information about each individual: (1) whether the individual is eligible for this measure (equivalently, whether the measure applies to this patient); and (2) a score for each eligible individual. For example, consider the following typical quality measure taken from the NQMC, one of 20 measures submitted by the National Diabetes Quality Improvement Alliance:

> Adult diabetes: percentage of patients with most-recent HbA1c level greater than 9.0 percent (poor control).

For this measure, the eligible individuals are patients diagnosed with diabetes, and the score for an eligible individual is 1 if the individual's most-recent HbA1c level exceeds 9 percent and 0 otherwise. To obtain the score for an eligible individual, one must scan historical laboratory results for each patient. Other measures may only require looking at the record for a single patient encounter.

Measures and Subsets of the Beneficiary Population

Measures can be specified in various ways to serve different purposes. If we include all DoD beneficiaries with diabetes in the denominator, we obtain a measure of overall MHS performance. If we include only beneficiaries treated at a specific MTF, then we will measure MTF performance. If we include only those treated by a particular physician, we are measuring physician performance. Similarly, we can define a measure for only active duty service members, or only females, or only beneficiaries between 45 and 75 years of age, or any combination of these conditions. Thus this single National Diabetes Quality Improvement Alliance "measure" is actually a template for a large number of specific measures, one for every possible subset of DoD beneficiaries diagnosed with diabetes.

There are many sensible ways to partition the population. For example, if older diabetics are more likely to have high HbA1c levels, it makes sense to define separate measures by age group, so that an MTF is not mistaken for a low performer simply because the diabetics it treats are older than the average for the whole MHS.

It also makes sense to look separately at beneficiaries who receive most of their care from MHS facilities (direct care) versus those who receive a large portion of their care from civilian providers (purchased care or non-MHS sources). AHLTA will provide medical record data for direct care (for ambulatory care at present; for all care once AHLTA is fully deployed). The MHS currently receives claims data for purchased care, and no data for care delivered by non-MHS sources. As explained in Chapter Five, MHS will then be able to calculate a richer set of measures for beneficiaries who mostly receive direct care.

Composite Measures

So far we have discussed only one of the 20 measures that the National Diabetes Quality Improvement Alliance submitted to the NQMC. If we define a set of measures for each one, as we did for the first, we multiply the total number of measures by 20. To mitigate this proliferation, we define *composite* measures, i.e., measures defined from other measures. Here are three of the National Diabetes Quality Improvement Alliance measures, including the one we have been discussing:

1. Adult diabetes: percentage of patients with most-recent HbA1c level greater than 9.0 percent (poor control).
2. Adult diabetes: percentage of patients with most-recent low-density lipoprotein–cholesterol (LDL-C) less than 130 mg/dL.
3. Adult diabetes: percentage of patients with most-recent blood pressure less than 140/90 mm Hg.

These three measures can be combined into a composite measure:

> Adult diabetes: percentage of patients with most-recent HbA1c level greater than 9.0 percent (poor control), or most-recent LDL-C greater than or equal to 130 mg/dL, or most-recent blood pressure greater than or equal to 140/90 mm Hg.

The denominator would be a subset of DoD beneficiaries with diabetes. The numerator would be a count of patients from the denominator who either (1) belong in the numerator of the first measure, or (2) *do not* belong in the numerator of the second measure, or (3) do not belong in the numerator of the third measure. The composite measure counts patients who exhibit unhealthy values for any of the three measures it was constructed from.

A composite measure summarizes the measures it is constructed from. One might collect all the detailed measures that address healthcare for diabetics and try to build a composite that measured the overall quality of care for diabetics. One could build such composite measures for asthma, chronic obstructive pulmonary disease (COPD), heart failure, and other chronic conditions, and build a composite of those composites that measures the overall quality of care for chronic conditions.

Assume one could map the detailed measures to the MHS strategic objectives. (The measure catalog described in the appendix contains such a mapping, but it is not unique. Many measures are mapped to two strategic objectives.) From the detailed measures associated with a particular strategic objective, one could build a composite that measured the degree to which the MHS was achieving that strategic objective.

If the composite measure indicates that all is well, one can ignore the components. If the composite measure indicates that there is a problem, one can drill down to the component measures to shed more light on the source and nature of the problem. Thus a composite measure should be designed to indicate as accurately as possible when the consumers of the measure—for example, the MHS senior leadership—need to look deeper. If the measure indicates that all is well when it is not, leadership will ignore problems that they should be addressing. If the measure indicates that problems exist when in fact they do not, the leadership will waste effort.

Composite Measures from Logic Models and Event Chains

Another way to reduce the number of measures to be considered involves logic models. We discussed logic models and their event chains in Chapter Three. They can be useful even if they contain no causal links to AHLTA.

Consider again the above HbA1c measure: Adult diabetes: percentage of patients with most-recent HbA1c level greater than 9.0 percent (poor control). The denominator contains a subset of DoD beneficiaries; the numerator, a subset of the denominator—namely, those with most-recent HbA1c level greater than 9 percent. We can further divide the patients who are in the denominator but not in the numerator into those who have a recent HbA1c level in their records and those who do not. The measure could be low for either of two reasons: (1) almost all diabetics have recent HbA1c levels and most of their levels are less than 9 percent; or (2) only a few diabetics have recent HbA1c levels. It is good if the measure is low for the first reason, but bad if it is low for the second reason.

The National Diabetes Quality Improvement Alliance submitted measures to deal with this issue.

Adult diabetes: percentage of patients receiving one or more HbA1c test(s).

Adult diabetes: percentage of patients with at least one LDL-C test.

There is no similar measure asking whether the patient's blood pressure has been measured.

We can think of a simple event chain consisting of two links: first, test a patient's HbA1c level; second, get the result of the test. We associate a measure with each link. The percentage of patients who are tested (first link) indicates whether the patient is receiving the proper care. The percentage of patients for whom the result exceeds 9 percent (second link) indicates something about how effectively the patient is controlling his or her blood glucose (the HbA1c level is an indicator of blood glucose control), which has implications for the likelihood of suffering microvascular complications of diabetes later on.

We could add a third link at the beginning of the chain, recommending that the patient be tested. The patient getting the test is now the second link, and obtaining the result of the test is the third link. Now we have three measures: (1) percentage of diabetic patients who receive a recommendation for a HbA1c test; (2) percentage of patients receiving the recommendation who actually get the test; and (3) percentage of patients who got the test that have a result over 9 percent. We may regard the first measure as indicating whether patients in the subset receive proper care, the second measure as indicating whether they are partnering with their providers, and the third measure as indicating whether they are successfully maintaining their health.

We can readily define an overall measure for either cause-and-effect chain. It is the percentage of patients with diabetes whose most-recent HbA1c level is either not known or exceeds 9 percent. If this percentage is low enough, there is no need to examine the measures individually. If the percentage is high enough to indicate a problem, then one can drill down to the component measures to shed light on the source and nature of the problem.

Selecting Measures to Display to Senior Leaders

We started with a large number of detailed measures (more precisely, measure templates), such as the National Diabetes Quality Improvement Alliance measure cited at the beginning of this section. We described how to generate even more measures from them by varying the subset of DoD beneficiaries on which a measure is defined. True, we can reduce the number of measures we need to look at by defining composite measures, but we must choose the subsets of beneficiaries for which we define them, as well. Even the thoughtful construction of composite measures can result in an impracticably large number of measures.

We think that the best way to deal with the problem of measure proliferation is to establish an analytical support group within the strategic management system. This group can produce the standard reports that managers will doubtless want to see, but they have two more important functions. First, they will screen the measures to select the information that senior leaders will need in order to decide where problems may exist and which problems they need to address first. They will design composite and aggregate measures[1] and nonstandard reports to convey this information economically. Second, if needed, they will drill down from the composite measures to their component measures as described earlier, to shed light on the source and nature of the problem.

Uses of Strategic Measurement

In this section, we discuss uses of strategic measures. First, they can alert senior leadership to problems. Second, they can help the leadership diagnose those problems and decide how to respond to them. Third, they can sometimes be part of that response.

Recognizing Problems

To recognize a problem, one needs a measure and something to compare it to. For example, is it a problem that 25 percent of diabetics have HbA1c levels above 9 percent?

One can compare this number to a benchmark, perhaps obtained from other healthcare systems. For this measure, the NCQA states that

---

[1]   The IOM (2006, p. 91–97) reports interest and some research into composite measures, but little actual use of them in civilian healthcare. We suggest that MHS analysts nevertheless review the literature for composite measures and borrow what they can.

commercial health plans reported in 2006 that 29.7 percent of diabetics had a most-recent HbA1c level over 9 percent. Medicare reported 23.6 percent, and Medicaid reported 49.1 percent (NCQA, 2006). In this context, 25 percent does not appear to be excessive.

One can also compare this number to the same measure prior to the implementation of AHLTA. Tracking a measure over time in this way can reveal trends, either worrisome or promising. Even if the measure is average or better today, a worrisome trend (for this measure, an upward trend) might signal a looming problem. Conversely, even if the measure is worse than average today, a promising trend may signal that the problem is on the way to being solved. It is unwise, perhaps, to accept this interpretation unless one understands why the trend is promising. For example, one may have done something in the past year to fix the problem and see the promising trend as evidence that the fix is working.

It is important to realize, however, that information is lost when measures are summarized, aggregated, or otherwise combined. Because of this loss of information, the aggregate measure can tell us something different than the detailed measures—it can mislead us. Depending on how the aggregate measure is constructed, it can tell us all is well when the detailed measures are showing problems, or it can indicate that there are big problems when in fact the problems are small.

A common way to construct an aggregate measure is to average many detailed measures. For example, the percentage of diabetics whose HbA1c exceeds 9 percent (poor control) can be calculated for the MHS as a whole, or for individual MTFs. The figure for the MHS as a whole can look entirely reasonable even though one or two MTFs have unacceptably high percentages. The overall average for the MHS is concealing a problem that the more-detailed MTF-specific measures identify.

This problem can be avoided by combining the detailed MTF-specific measures in a different way—for example, by defining the aggregate measure as the maximum percentage observed at any MTF. This approach tends to amplify the apparent problem because the measure has the same value regardless of whether all MTFs have high percentages or only one.

Intermediate kinds of measures are also possible, measures that are less likely than the simple average to conceal problems but also less likely than the maximum over MTFs to overstate problems. But no aggregate measure avoids both errors altogether.

Variation is often a signal of problems. One finds variation by comparing how different entities within the MHS use AHLTA to manage diabetic patients. If very different percentages of diabetics have HbA1c levels over 9 percent, depending on which MTF provides their treatment, then perhaps clinicians in low-performing MTFs can learn better approaches from clinicians in high-performing MTFs. One must be careful, of course, to adjust for sources of variation other than the MTF, such as case-mix differences.

It is necessary, then, to use good judgment in the construction and use of aggregate measures. It is reasonable, for example, to aggregate a set of detailed measures in several ways, each with a different propensity to overstate or understate problems. If no problem is shown by an aggregate measure that tends to overstate problems, one probably needn't look any deeper. An aggregate measure thus becomes a screen to focus a manager's attention where it most needs to be.

Diagnosing Problems

Earlier we discussed one strategy for diagnosing a problem once it has been recognized—namely, drilling down from composite measures to their component measures. But any technique for analyzing the data available to the MHS could contribute.

Recall that each detailed measure serves as a template for innumerable strategic measures, with each strategic measure obtained by specifying a subset of DoD beneficiaries. The subset may be defined using any data elements that can be associated with each beneficiary that might potentially contribute to the value of the measure. Data elements could include demographic factors (the patient's age, gender, ethnicity), socioeconomic factors (income, type of job), and factors related to health and healthcare (treating MTF, comorbidities).

It is worthwhile to define standard subsets, such as patients by age and treating MTF. Measures based on standard subsets are useful for tracking performance changes over time. But it is also worthwhile to build models that estimate the value of a measure as a function of the various data elements.

Fixing Problems

Strategic measurement of performance will not fix problems in and of itself, but it can play a role. First, one can publish performance differences between MHS entities, such as MTFs or individual clinicians, and map them to differences in the use of AHLTA. This can provide managers and individuals motivational feedback for the use of AHLTA in performance improvement.

Second, process improvement projects in healthcare (and other industries) generally utilize an iterative process of measurement, change, and measurement. The analysis done to diagnose a problem yields a hypothesis about what can be done to fix it. As part of implementing the fix, one should track whether it is having the desired effect and if not, adjust it. For example, one might measure the compliance rate with preventative measures before and after a reminder system is installed to tailor and fine-tune the reminders.

CHAPTER SEVEN

Concluding Comments

The objective of this project was to help the MHS develop a framework and define specific measures to be used to judge and report the efficiency, safety, and health benefits of the AHLTA electronic medical record system as it becomes fully deployed. In this chapter, we briefly summarize the framework we developed and discuss what remains to be done to implement the measures and use them to assess AHLTA's contribution to MHS performance.

Framework

The framework consists of three elements:

1. The MHS strategic objectives, which are the dimensions of performance along which AHLTA should be judged.
2. Our recommended approach for estimating the effects of specific uses of AHLTA on the basic measures. Previously, we called the specific uses of AHLTA "treatments." Our approach lays out a logic model that describes a causal chain connecting the treatment to the basic measures. To estimate the overall effects of AHLTA on the high level measures requires that we aggregate or summarize not only over basic measures, but over treatments as well.
3. Specific outcome measures or measurement approaches for the purpose of assessing AHLTA's effect on MHS performance. The raw data for calculating measures describe the MHS one patient

and one encounter at a time. The basic measures are calculated at this very detailed level, and higher-level measures are formed by selecting, aggregating, and summarizing the detailed measures. High-level managers need high-level measures to avoid information overload, but when a high-level measure shows that a problem exists, it must be possible to drill back down to more detailed levels in order to pinpoint that problem.

Implementing Measures

Three conditions must be satisfied in order to implement the measures we suggest in Chapter Five. First, one must define an algorithm for calculating it. The algorithm identifies all the data elements that appear in the calculation and gives the formula or procedure for combining the data elements into the measure. Algorithms have been defined for all the NQMC measures and will be simple to devise for IMR measures. The MHS has developed algorithms for the PMPM efficiency measures. Algorithms are lacking for other measures we discussed in Chapter Five (e.g., for medical care during military operations or for public health surveillance).

Second, a measure should be tested for reliability and validity. According to the NQMC Web site, reliability is the degree to which the measure is free from random error, while validity is the degree to which the measure is associated with what it purports to measure.[1] Submitters of measures to the NQMC must provide documented peer-reviewed evidence evaluating the reliability and validity of the measures they submit. Until algorithms have been constructed for other measures discussed in Chapter Five, it will not be possible to assess their reliability and validity.

Third, one must identify sources for the needed data elements, a task that remains to be done. Can a data element be found in existing MHS data systems, such as the Standard Inpatient Data Record (SIDR) or SADR? Will it be captured in AHLTA's CDR, and, if so,

---

[1]   See National Quality Measures Clearinghouse (NQMC), 2007.

will it truly be a data element or will it be buried in free text? It is also important to know whether the data elements will be accurately and consistently available. If a data element is missing much of the time, or is interpreted differently at one installation than another, measures calculated using the data element may be meaningless noise.

Technical difficulties may arise when a measure requires information from multiple sources. For example, the PMPM efficiency measures require data on cost (from MEPRS) and on work done (from the SIDR and SADR, for example). An algorithm had to be constructed to allocate cost (in the categories used in MEPRS) to units of work (in the categories used in the SIDR and SADR).

Measures must be implemented in a way that accommodates beneficiaries with missing data. As AHLTA and the CDM/CDW are deployed, they will make progressively more data available on a substantial fraction of MHS beneficiaries, but data will be collected for some but not all beneficiaries. The data available for a beneficiary will be most complete if the beneficiary received all his care from the direct care system, and less complete if he receives some or all his care from the purchased care system or out of the system.

There are several ways to accommodate missing data. For example, we may identify groups of beneficiaries with much the same data elements available—for example, those whose primary care physician is in the direct care system, versus in the purchased care system, versus neither. We might then apply different measures to the different groups. Or we can revise a measure to accommodate missing data. Recall that a measure is built up from two items of information about each individual: (1) whether the individual is eligible for this measure (equivalently, whether the measure applies to this patient); and (2) a score for each eligible individual. One can accommodate missing data by adjusting either item of information. Thus one can redefine eligibility for the measure to exclude those for whom critical data elements are missing, or one can add a score that means "data are missing."

Logically, it should be possible to make better, more-reliable measurements of MHS performance with more-complete data. Since the deployment of AHLTA and the CDM/CDW will make more data available, it should be possible to improve these measurements. However, it is beyond the scope of this project to consider this issue in depth.

Analysis Capability

The framework requires an entity that can calculate the measures and perform the analyses needed to assess the benefits of AHLTA. The entity that performs analytic functions will need ready access to the CDM/CDW and many other data systems, including those that are collected and maintained by the Executive Information and Decision Support (EIDS) system.[2]

Much more will be required of this entity than simply assembling and cleaning data, calculating sets of measures, and generating a standard set of reports each month. It must also perform analyses using the methods from Chapter Three. As pointed out in Chapter Five, the currently available measures are not sufficient for the needs of the MHS. Additional detailed measures should be developed to cover strategic objectives whose current coverage is deficient. Strategic measures must be constructed from the detailed measures. The organizational home of the framework should be involved in these tasks, though not necessarily without help.

We anticipate that many of the treatments to be analyzed will originate as process improvement exercises at individual installations. The entity will need to provide analytic support for these efforts.

If the MHS as a whole is to benefit from an improved process developed at one installation, the improvement must be spread to others. The analysis organization can act as a clearinghouse for information on all installations' improvement efforts. This is by no means sufficient to spread improvements, but it may help.

Finally, the measures we have discussed capture progress toward the MHS strategic objectives from any cause, not just from the use of AHLTA. Ultimately, the organization selected to host the framework

---

[2]   See the EIDS home page (MHS, 2007).

should not focus exclusively on estimating benefits from AHLTA, although this may be its initial focus. It should look ahead to providing reports and analysis of broader utility.

Catalog of NQMC Measures

We have constructed a catalog of all 1,153 measures available on the NQMC Web site as of April 11, 2007. (See NQMC, 2007). It is located on the Web at:

http://www.rand.org/pubs/monographs/MG680/

It is in the form of a Microsoft Excel file, "Catalog of NQMC Measures 070411.xls," with the following data elements:

1. NQMC number
2. Title
3. Preferred result
4. Denominator and numerator
5. Measure source
6. Measure group and subgroup
7. Primary and secondary NQMC domains
8. 1st through 3rd IOM aims
9. 1st through 4th IOM need
10. 1st through 4th data source
11. Current and future computability
12. Useful measure of AHLTA's effect? Why or why not?
13. 1st and 2nd strategic objective

We obtained data for items 1–10 from the "complete summaries" of the measures, which we downloaded from the NQMC site between December 20, 2006 and April 11, 2007. Some of the summaries may have been revised since then. We generated the information for items 11–13 as described below.

NQMC Number

NQMC has given each measure a six-digit number, which we have included in the catalog as a convenient way to identify a measure on the NQMC site. In the catalog, this data element takes the form "NQMC123456."

Title

We copied the title of each measure verbatim from the complete summary.

Preferred Result

This data element is a slightly edited version of an item called the "Interpretation of Score" in each complete summary. Its possible values are shown in Table A.1.

Table A.1
Possible Values of "Preferred Result"

Value in "Preferred Result" Column	Count
Better quality is associated with a higher score,	832
Better quality is associated with a lower score,	135
Better quality is associated with a defined interval,	2
Passing score defines better quality,	16
Unspecified	168
Total	1,153

Denominator and Numerator

Virtually all the NQMC measures are ratios. Each measure is constructed by selecting occurrences of some set of conditions from a database, and giving each occurrence a score. The denominator is a count of the occurrences; the numerator is the sum of the scores of the occurrences in the denominator.

Measure NQMC000596, "Adult diabetes: percentage of patients receiving one or more A1c test(s)," provides a particularly simple example. The denominator is defined as "All patients diagnosed with diabetes aged 18–75 years," and the numerator as "The number of patients from the denominator who received one or more A1c test(s)." A more-complex example might restrict the denominator to people who have been members of a particular health plan for at least a specified period of time and might restrict the numerator to people whose most-recent A1c test was performed within a specified time window.

The catalog descriptions of the denominators and numerators are edited versions of the descriptions in the complete summaries.

Measure Source

As described in Chapter Five, AHRQ invites developers of quality measures to submit measures that meet certain inclusion criteria. Table A.2 shows the organizations that appear in the catalog as the sources of measures. The complete summaries sometimes name multiple sources for a measure. For the catalog we arbitrarily selected one of them.

Table A.2
Source Organizations of NQMC Measures

Source Organization	No. of NQMC Measures
Accreditation Association for Ambulatory Health Care Institute for Quality Improvement	2
AHRQ	121
American Medical Directors Association	65
Arthritis Foundation	34
British Medical Association/National Health System (NHS) Confederation	56
Canadian Cardiovascular Outcomes Research Team	19
Child and Adolescent Health Measurement Initiative	7
Child Health Corporation of America	3

Table A.2—continued

Source Organization	No. of NQMC Measures
Centers for Medicare and Medicaid Services (CMS)	78
Family Violence Prevention Fund	8
Focus On Therapeutic Outcomes, Inc	6
HealthPartners	2
HEDIS 2006	62
HRSA	42
ICSI	117
Inouye, Sharon K. M.D., M.P.H.	1
Joint Commission	38
Kolcaba, Katharine Ph.D.	1
Manitoba Centre for Health Policy	13
McLean Hospital	7
National Diabetes Quality Improvement Alliance	20
New York State Department of Health AIDS Institute	175
Physician Consortium for Performance Improvement	53
Press Ganey Associates, Inc	63
Renal Physicians Association	35
Therapeutic Associates, Inc.	1
VHA	43
VHA Mental Illness Research, Education and Clinical Center (MIRECC)	18
Wisconsin Collaborative for Healthcare Quality	2
Wisconsin DHFS	61
Total	1,153

Measure Group and Subgroup

On the NQMC site, measures from each source are organized into a hierarchy. The hierarchy containing a measure may consist of as many as five levels, but it usually consists of no more than three (the measure and two higher levels) and often no more than two. The hierarchies for different sources do not appear to have a common organizing principle.

We selected up to two levels above the measure as our group and subgroup.

- For measures with no levels between the measure and the source (i.e., intermediate levels), we left the group and subgroup blank.
- For measures with one intermediate level, we set the group to the name of that level and left the subgroup blank.
- For measures with two intermediate levels, we set the group to the higher of the levels and the subgroup to the lower.
- For measures with more than two intermediate levels, we chose the group and subgroup to be two of the levels. To the degree possible, we chose the subgroup to be a disease or condition, such as asthma, diabetes, or acute myocardial infarction, and the group to be the next higher level.

In Chapter Six, we discussed the need to construct strategic (i.e., high-level) measures from detailed measures. A reasonable starting point, we think, is with a set of measures that have the same source, group, and subgroup. One could construct a single strategic measure that summarizes the entire set.

Primary and Secondary NQMC Domains

Each NQMC measure is assigned to one or two of seven domains. If a measure is assigned to two domains, one is designated as the primary domain. The NQMC describes the domains as follows:

1. **Access:** a patient's or enrollee's attainment of timely and appropriate healthcare.
2. **Outcome:** health state of a patient resulting from healthcare.
3. **Patient Experience:** a patient's or enrollee's report concerning observations of and participation in healthcare.

4. **Population Health:** the state of health of a group of persons defined by geographic location, organizational affiliation or nonclinical characteristics. (Eligibility for measures of population health is not restricted to recipients of clinical care.)

5. **Process:** a healthcare service provided to, on behalf of, or by a patient that is appropriately based on scientific evidence of efficacy or effectiveness.

6. **Structure:** a feature of a healthcare organization or clinician relevant to its capacity to provide healthcare.

7. **Use of Services:** the provision of a service to, on behalf of, or by a group of persons defined by geographic location, organizational or nonclinical characteristics without determination of the appropriateness of the service for the specified individuals. Use of service measures can assess encounters, tests, interventions as well as the efficiency of the delivery of these services.

The NQMC prefers measures to be assigned to access, outcome, patient experience, process, or structure. It makes the other two domains (population health and use of services) available for measures used in conjunction with clinical performance measures as part of a measure set. Table A.3 gives the number of measures assigned to each domain.

Table A.3
Number of Measures by NQMC Domain Assignment

NQMC Domain	Primary	Secondary
Access	22	56
Outcome	193	8
Patient Experience	247	10
Population Health	29	
Process	597	34
Structure	41	
Use of Services	24	
Total	1,153	108

We included the NQMC domain assignments in the catalog because we expected it would help us assign measures to the MHS strategic objectives (discussed below).

1st, 2nd and 3rd IOM Aims

The NQMC has adopted the six aims of healthcare quality proposed in 2001 by the IOM (2001a):

- **Effectiveness:** Provide care processes and achieve outcomes as supported by scientific evidence.
- **Efficiency:** Avoid waste, including waste of equipment, supplies, ideas, and energy.
- **Equity:** Provide care that does not vary in quality because of personal characteristics such as gender, ethnicity, geographic location, and socioeconomic status.
- **Patient centeredness:** Meet patient's needs and preferences and provide education and support.
- **Safety:** Avoid actual or potential bodily harm.
- **Timeliness:** Minimize delays to obtaining needed care.

The complete summary for each measure lists the zero (52 measures) to three (39 measures) IOM aims to which it applies. Table A.4 shows the number of measures assigned to each IOM aim.

Table A.4
Counts of Measures by IOM Aim
Assignment

IOM Aim	No. of Measures
Effectiveness	805
Efficiency	0
Equity	42
Patient centeredness	368
Safety	64
Timeliness	77

We included the IOM aim assignments in the catalog because they form one of the dimensions of the classification used by the National Healthcare Quality Report (NHQR) (AHRQ, 2006; IOM, 2001b). They also appear prominently in the descriptions of the MHS strategic objectives IP1 (evidence based medicine is used to improve quality, safety, and appropriate utilization of services) and IP3 (our healthcare processes are patient centered, safe, effective, and efficient). We anticipated, therefore, that these data elements would help us assign the NQMC measures to strategic objectives (see below).

1st, 2nd, 3rd and 4th IOM Need

The NQMC measures are also assigned to four consumer perspectives on healthcare needs described in "Envisioning the National Health Care Quality Report" (IOM, 2001b). These perspectives are:

- **End of life care:** Care related to those not expected to survive more than six months.
- **Getting better:** Care related to acute illness or injury.
- **Living with illness:** Care related to chronic or recurrent illness.
- **Staying healthy:** Care related to healthy populations or the general health needs of nonhealthy populations (e.g., health promotion, disease prevention, risk factor assessment, early detection by screening and treatment of pre-symptomatic disease).

The complete summary for each measure lists the zero (52 measures) to four (29 measures) IOM care needs to which it applies. Table A.5 shows the number of measures assigned to each IOM care need.

We included the IOM care need assignments in the catalog because they form one of the dimensions for classifying measures used in the NHQR (AHRQ, 2006; IOM, 2001b) and in the recent IOM report on healthcare performance measurement (IOM, 2006).

Table A.5
Counts of Measures by IOM
Care Need Assignment

IOM Care Need	No. of Measures
End of life care	84
Getting better	378
Living with illness	785
Staying healthy	177

1st, 2nd, 3rd and 4th Data Source

The complete summary for each measure identifies the types of data needed to compute the measure. Measures require from one (653 measures) to four (32 measures) types of data. Table A.6 shows the number of measures that require data of each type. We included these data elements in the catalog so that we can crudely estimate whether a measure can be calculated from data currently available to the MHS, whether

Table A.6
Types of Data Needed to Compute
Measures

Data Source	No. of Measures
Administrative	607
Clinician survey	19
Laboratory	85
Medical record	536
Patient survey	305
Pharmacy	57
Population survey	1
Provider	10
Public Health	9
Registry	87
Special or unique	83

it can be calculated once AHLTA has been fully implemented, or whether calculating it requires a type of data (e.g., surveys) that we do not expect will be routinely available either at present or once AHLTA is fully implemented.

Current and Future Computability

We used the types of data required to compute each measure to estimate whether the measure could be computed under either of two circumstances:

- At present with routinely available data, such as the data available in the MHS Data Repository (MDR)
- In the future, once AHLTA has been fully implemented.

Our rules for making these determinations are simple. We assumed that administrative, pharmacy, and public health data are routinely available today, so if a measure needs no other data types it can be computed today. We assumed that laboratory, medical record, and registry data will also be routinely available wherever AHLTA is implemented, so measures that need no more these six types of data will become computable. There are no plans to implement AHLTA in long-term care settings or home health agencies, so the computability of measures of care provided in these settings will not be affected by AHLTA's implementation. Measures that require any of the remaining data types will not be routinely computable even after AHLTA is implemented. Table A.7 shows the counts of measures by present and future computability status.

Table A.7
Computability Status of Measures

Computability	Present	Future
No	975	457
Yes	178	696

We stress that this is only a crude determination of computability. Each measure requires very specific data elements. Having access to all the required data types does not guarantee that one has all the required data elements. In particular, EMRs are not standardized. One EMR may store an item of information in coded form (as a machine-interpretable data element); another may store the same information embedded in text. In the second case, we would not consider the measure to be computable on a routine basis.

In addition, different data elements will be available for different DoD beneficiaries. The MHS provides healthcare both from its own facilities (direct care) and by contracting with civilian providers (purchased care). Under current plans, AHLTA will be used in all MHS facilities but not by the providers of purchased care. Active duty service members receive the bulk of their healthcare from MHS facilities, and thus AHLTA data will be available for them. But this is less true of other DoD beneficiaries.

Useful Measure of AHLTA's Effect? Why or Why Not?

We created a data element in the catalog that expresses our tentative answer to the question: "Is there a treatment—a specific way of using one or more of AHLTA's features and functions under defined circumstances—whose influence on this NQMC measure can be estimated using the methods described in Chapter Three?" A second data element gives a very brief reason for that answer.

We judged that our method might be used to estimate the effect of some treatment on 629 of the NQMC measures, but not the other 524. When we answered "Yes," the reason we gave suggested how AHLTA might influence the measure:

- Many measures can be influenced through decision support at the point of care (i.e., reminders, warnings, alerts, and guidelines).
- Measures associated with chronic conditions or diseases (e.g., diabetes, asthma, chronic kidney disease, HIV/AIDS, and depression) can be influenced through disease management programs.

- Some measures could be influenced through process improvement efforts using data collected in AHLTA's CDW/CDM.

Our most common reasons for answering "No" were the following:

- The measure has an unspecified preferred result, so influencing it has no value.
- The measure is designed to assess care in settings where AHLTA will not be implemented, such as long-term care and home health care.
- Calculating the measure requires a type of data that we do not think will be routinely available, such as surveys or "special or unique" data (see Table A.6).
- The measure is not focused enough, meaning that we expect so many factors to influence the measure that isolating AHLTA's effect is hopeless. Measures with a primary NQMC domain of Population Health all fall into this category.

1st and 2nd Strategic Objective

We tentatively assigned the NQMC measures to strategic objectives. Here we explain the reasoning behind our assignments, but we emphasize that we do not consider the result to be entirely satisfactory. The reader should take our assignments as suggestions only, and should feel free to change them to conform to his own judgment.

As described in Chapter Five, we associate the NQMC measures with the first theme from the MHS strategic plan (OASD, 2007a), "manage and deliver the benefit." With very few exceptions, we have assigned the NQMC measures only to strategic objectives associated with this theme (Table A.8).

We developed the following rules of thumb to guide our assignment. First, we did not assign measures with an unspecified preferred result to any strategic objective. There is a preferred result for a strategic objective, namely, to approach or achieve it. We reasoned that a

Table A.8

**Strategic Objectives Associated with Theme I:
Manage and Deliver the Benefit**

Index	Objective
S3	Beneficiaries are satisfied with their healthcare.
S4	The MHS creates healthy communities.
F1	DoD healthcare costs are managed; benefit is sustained and shaped.
C1	"I am a partner with my healthcare team. We know and care about improving my health."
C2	"It feels like the MHS was designed just for me."
IP1	Evidence-based medicine is used to improve quality, safety, and appropriate utilization of services.
IP2	Beneficiaries partner with us to improve health outcomes.
IP3	Our healthcare processes are patient centered, safe, effective, and efficient.
IP6	Comprehensive globally accessible health and business information enables medical surveillance, evidence based medicine, and effective healthcare operations.

measure with no preferred result probably will not contribute to the assessment of a strategic objective.

We based the assignment of measures with any other preferred result on the primary NQMC domain as follows:

Access, as defined by the NQMC, is a patient's or enrollee's attainment of timely and appropriate healthcare. The very word "access" suggests patient centeredness, and each of these measures had either "Effectiveness" or "Patient Centeredness" as an IOM domain. We therefore assigned all measures with a primary NQMC domain of Access to strategic objective IP3.

Outcome is a health state of a patient resulting from healthcare. However, one must interpret "health state" liberally. Death is obviously an outcome, and so mortality rates are outcome measures. But the NQMC also considers a diabetic's hemoglobin A1c level or a hypertensive patient's blood pressure to be an outcome, and so the percentage of diabetics with HbA1c levels above 9 percent (poor control) and the percentage of hypertensive patients with a blood pressure under 140/90 (good control) are also outcome measures.

With four exceptions, we assigned measures with a primary NQMC domain of Outcome to both strategic objectives S4 or IP3. The appropriate assignment is S4 when outcome measures are used to assess the health of a community. The appropriate assignment is IP3 when they are used to make inferences about the effectiveness of care provided by individual providers.

Three of the four exceptions involved measures with a denominator that included all people eligible for a test and a numerator that included all people with a given result for the test. Thus they combined the questions of whether a person eligible for a test was actually tested, and whether therapy was having its desired effect. Our rules of thumb assign measures that ask the first question to IP1, and measures that ask the second to IP3. Therefore we assigned these three measures to IP1 and IP3.

The remaining outcome measure asked for the percentage of women that had completed their cervical cancer screening within six months of a reminder sent to them. This seemed to address the issue of beneficiaries partnering with their providers, and so we assigned it to both C1 and IP2.

Patient Experience is a patient's or enrollee's report concerning observations of and participation in healthcare. We assigned all measures with this primary NQMC domain to two strategic objectives. First, we assigned them all to S3, "Beneficiaries are satisfied with their healthcare."

But we also felt that patient experience measures ought to be assigned to one of the customer perspective strategic objectives. If the measure title mentioned counseling or education or suggested active patient involvement, we secondarily assigned it to C1, "I am a partner with my healthcare team; we know and care about improving my health." Otherwise, its secondary assignment was C2, "It feels like the MHS was designed just for me."

Population Health. We assigned all measures with a primary NQMC domain of Population Health to strategic objective S4, "The MHS creates healthy communities."

Process is defined by the NQMC as a healthcare service provided to, on behalf of, or by a patient appropriately based on scientific evidence of efficacy or effectiveness. We somewhat arbitrarily equated NQMC's notion of "process" with the "internal process" perspective of the MHS strategy map, and restricted our assignments of process measures to objectives IP1, IP2, or IP3 (IP6 was not a plausible assignment for any of the process measures).

We assigned a measure to IP3 if

- it addressed the safety or effectiveness of care
- calculating the measure required the result of a test
- it addressed patient centeredness
- it mentioned counseling or education of patients, or discussion between patients and providers, or instructions given to patients
- it could be interpreted as measuring whether something was done right.

We assigned a measure to IP2 if

- the measure description mentioned self-care
- the measure mentioned a patient actually doing something about his health, e.g., actually exercising (as opposed to being advised to exercise) or actually filling a prescription (as opposed to the provider writing a prescription)
- there was any other suggestion in the measure description that the patient had taken an active role in his own care.

We assigned a measure to IP1 if

- it addressed the appropriateness of a service (i.e., whether the right thing was done)
- it mentioned evidence-based medicine or guidelines
- calculating the measure required knowing whether a test was done, but not the result of the test
- neither IP2 nor IP3 applied (i.e., IP1 was the default assignment).

Table A.9 shows the counts of process measures assigned to each strategic objective. Two measures were assigned to both IP2 and IP3.

Table A.9
Counts of Process Measures Assigned by Strategic Objective

Strategic Objective	No. of Measures
IP1	478
IP2	13
IP3	108
Total	599

These measures counted schizophrenic patients who had been out of contact with care providers for an extended period. We reasoned that they deserved both assignments because either the provider or the patient (or the patient's family) could influence these measures.

Structure. Fifteen measures with this primary NQMC domain asked whether a practice can produce a registry of patients. We assigned them to IP6, "Comprehensive globally accessible health and business information enables medical surveillance, evidence based medicine and effective healthcare operations," which we interpret to be about the information infrastructure of the MHS.

Sixteen structure measures assess provider competence or training. We assigned them to strategic objective L5, "Personnel are recruited, trained, educated, and retained to meet requirements."

Six structure measures use volumes of specified services (e.g., number of coronary artery bypass grafts a hospital performs annually) as indirect indicators of quality. These measures are based on the finding that hospitals with high volumes generally achieve better results. We assigned them to IP3, "Our healthcare processes are patient centered, safe, effective, and efficient."

Two structure measures dealt with waiting times for appointments. We treated them like access measures and assigned them to IP3.

Of the remaining two structure measures, we assigned one (Does the provider comply with a protocol?) to IP1 and one (Does the practice staff regularly discuss patients in its palliative care registry?) to IP3.

Use of Services. We did not assign measures with this primary NQMC domain to any strategic objectives. All of them had unspecified preferred results.

Bibliography

Academy Health (2006). *Efficiency In Health Care: What Does It Mean? How Is It Measured? How Can It Be Used for Value-Based Purchasing? Highlights from a National Conference.* As of May 6, 2007:
http://www.academyhealth.org/publications/EfficiencyReport.pdf

Agency for Healthcare Research and Quality (AHRQ) (2006). *2006 National Healthcare Quality Report.* As of May 28, 2007:
http://www.ahrq.gov/qual/nhqr06/nhqr06.htm

Air Force Surgeon General's Executive Global Look, October 2007. As of November 26, 2007:
https://sgegl.afmoa.af.mil/

Asch SM, McGlynn EA, Hogan MM, Hayward RA, Shekelle P, Rubenstein L, Keesey J, Adams J, Kerr EA (2004). Comparison of quality of care for patients in the Veterans Health Administration and patients in a national sample, *Ann Intern Med 141*(12):938–945.

Assistant Secretary of Defense, Health Affairs (ASD/HA) (2002). *Department of Defense Medicare Eligible Retiree Health Care Fund Operations,* Directive Number 6070.2, July 19, 2002.

Bigelow JH, Fonkych K, Fung C, Wang J (2005). *Analysis of Healthcare Interventions That Change Patient Trajectories,* Santa Monica Calif.: RAND Corporation, MG-408-HLTH. As of November 14, 2007:
http://www.rand.org/pubs/monographs/MG408/

Bodenheimer T, Wagner EH, Grumback K (2002). Improving primary care for patients with chronic illness, *JAMA* 288(14):1775–1779.

Chaudhry B, Wang J, Wu S, Maglione M, Mojica WA, Roth EA, Morton SC, Shekelle PG (2006). Systematic review: Impact of health information technology on quality, efficiency, and costs of medical care, *Ann Intern Med 144*(10):E12–E22.

CMS—*See* Command Management System.

Command Management System (2007). Password-protected Army Medical Department site. As of November 26, 2007:
https://sso.mods.army.mil/cms/secured/

Department of Defense, Office of the Inspector General (2006). *Information Technology Management: Acquisition of the Armed Forces Health Longitudinal Technology Application*, D-2006-089, May 18, 2006. As of May 9, 2007:
http://www.dodig.osd.mil/Audit/reports/FY06/06-089.pdf

Demakis JG, McQueen L, Kizer KW, Feussner JR (2000). Quality Enhancement Research Initiative (QUERI): A collaboration between research and clinical practice, *Medical Care* 38(6) QUERI Supplement:I17–I25.

Girosi F, Meili R, Scoville R (2005). *Extrapolating Evidence of Health Information Technology Savings and Costs* (2005). Santa Monica, Calif.: RAND Corporation, MG-410-HLTH. As of November 14, 2007:
http://www.rand.org/pubs/monographs/MG410/

Government Accountability Office (1998). *Executive Guide: Measuring Performance and Demonstrating Results of Information Technology Investments*, GAO/AIMD-98-89. As of April 24, 2007:
http://www.gao.gov/special.pubs/ai98089.pdf

Greenfield V, Williams VL, Eiseman E (2006). *Using Logic Models for Strategic Planning and Evaluation: Application to the National Center for Injury Prevention and Control*, Santa Monica, Calif.: RAND Corporation, TR-370-NCIPC. As of November 14, 2007:
http://www.rand.org/pubs/technical_reports/TR370/

Haynes RB, McDonald HP, Garg AX (2002). Helping patients follow prescribed treatment, *JAMA* 288(22):2880–2883.

Hillestad R, Bigelow J, Bower A, Girosi F, Meili R, Scoville R, Taylor, R (2005). Can electronic medical record systems transform health care? Potential health benefits, savings, and costs, *Health Affairs* 24(5):1103–1117.

Hsiao WC, Braun P, Dunn D, Becker ER, DeNicola M, Ketcham TR (1988). Results and policy implications of the resource-based relative-value study, *New England Journal of Medicine*, September 29, 1988; 319(13):881–888.

Hynes DM, Cowper D, Kerr M, Kubal J, Murphy PA (2000). Database and informatics support for QUERI: Current systems and future needs, *Medical Care* 38(6) QUERI Supplement:I114–I128.

Ingenix (Intelligence for Health Care) Web site (2007). As of November 13, 2007:
http://www.ingenix.com

Institute of Medicine (IOM) (2001a). *Crossing the Quality Chasm: A New Health System for the 21st Century*, Washington D.C.: National Academy Press.

——— (2001b). *Envisioning the National Health Care Quality Report*, Washington, D.C.: National Academy Press.

——— (2006). *Performance Measurement: Accelerating Improvement*, Washington. D.C.: National Academy Press.

Jha AK, Perlin JB, Kizer KW, Dudley RA (2003). Effect of the transformation of the Veterans Affairs health care system on the quality of care, *New England Journal of Medicine* 348(22):2218–2227.

The Joint Commission (2007). *Performance Measurement*. As of November 9, 2007:
http://www.jointcommission.org/PerformanceMeasurement/

Kilo CM (2005). Transforming care: Medical practice design and information technology, *Health Affairs* 24(5):1296–1301.

Kizer KW, Demakis JG, Feussner JR (2000). Reinventing VA Health Care: Systematizing quality improvement and quality innovation, *Medical Care* 38(6) QUERI Supplement:I7–I16.

Lohr KN, ed. (1990). *Medicare: A Strategy for Quality Assurance, Vol. I*, Washington, D.C.: National Academy Press.

McGlynn EA (2003). Selecting common measures of quality and system performance, *Medical Care* 41(1 Suppl):I39–I47.

McGlynn EA, Asch SM, Adams J, Keesey J, Hicks J, DeCristofaro A, Kerr EA (2003). The quality of health care delivered to adults in the United States, *New England Journal of Medicine* 348(26):2635–2645.

MHS—*See* Military Health System.

Military Health System (MHS) (2007). Executive Information and Support Home Page. As of November 14, 2007:
http://www.ha.osd.mil/peo/eids/default.asp

——— Conference (2006). *Developing a Resource Based Relative Value Scale for MHS Mission Essential Activities Not Currently Receiving RVU Value*, Presentation, Wednesday, February 1, 2006. As of May 7, 2007:
http://www.tricare.mil/conferences/2006/download/049Dinneen.ppt

——— Conference (2007a). *AHLTA User Outreach*, Presentation, Monday, January 29, 2007. As of April 30, 2007:
http://www.tricare.mil/conferences/2007/Mon/M303.ppt

——— Conference (2007b). *ESSENCE: Biosurveillance Enters the Age of Empirical Analysis*, Presentation, Monday, January 29, 2007. As of April 30, 2007:
http://www.tricare.mil/conferences/2007/Mon/M304.ppt

——————— Conference (2007c). *Military Health System (MHS) Metrics Standardization-Update*, Presentation, Monday, January 29, 2007. As of April 30, 2007:
http://www.tricare.mil/conferences/2007/Mon/M408.ppt

——————— Conference (2007d). *Application of Per Member Per Month (PMPM): Translating Strategy to Action at the Customer Interface*, Presentation, Wednesday, January 31, 2007, As of April 30, 2007:
http://www.tricare.mil/conferences/2007/Wed/W418.ppt

——————— Conference (2007e). *Per Member Per Month (PMPM): Metric Methodologies,* Presentation, Wednesday, January 31, 2007, As of April 30, 2007:
http://www.tricare.mil/conferences/2007/Wed/W420.ppt

——————— Conference (2007f). Awards/Administrative Announcements. As of November 9, 2007:
http://www.tricare.mil/conferences/2007/AgendaWed.cfm

Miller NH (1997). Compliance with treatment regimens in chronic asymptomatic diseases, *Am J Med* 102(2A):43–49.

Miller RH, Sim I (2004). Physicians' use of electronic medical records: Barriers and solutions, *Health Affairs* 22(2):116–126.

Mohr LB (1988). *Impact Analysis for Program Evaluation.* Chicago, Ill.: The Dorsey Press.

National Committee for Quality Assurance (NCQA) (2006). *The State of Health Care Quality: Industry Trends and Analysis*, Washington, D.C.: NCQA. As of May 15, 2007:
http://www.ncqa.org/communications/sohc2006/sohc_2006.pdf

——————— (2007). HEDIS Publications. As of November 9, 2007:
http://web.ncqa.org/tabid/78/Default.aspx

National Quality Measures Clearinghouse (NQMC), 2007. As of November 11, 2007:
http://www.qualitymeasures.ahrq.gov/

NQMC—*See* National Quality Measures Clearinghouse.

Nutting PA, Goodwin MA, Flocke SA, Zyzanski SJ, Stange KC (2003). Continuity of primary care: To whom does it matter and when? *Ann Fam Med* 1(3):149–155. As of April 25, 2007:
http://www.annfammed.org/

OASD (Health Affairs) (2007a). *Military Health System Strategic Plan.* As of April 27, 2007:
http://www.ha.osd.mil/strat_plan/MHS_Strategic_Plan_07Apr.pdf

———— (2007b). *MHS Balanced Scorecard*. As of April 27, 2007:
http://www.ha.osd.mil/strat_plan/

OIG—*See* Department of Defense, Office of the Inspector General.

Office of Management and Budget (1997). *Capital Programming Guide, Supplement to Part 7 of Circular No. A-11*. As of November 29, 2007:
http://www.whitehouse.gov/omb/circulars/a11/cpgtoc.html

Ornstein C (2003). Hospital heeds doctors, suspends software, *Los Angeles Times*, January 23, 2003.

Quadrennial Defense Review (QDR) (2006), *Roadmap for Medical Transformation*. As of April 27, 2007:
http://www.ha.osd.mil/strat_plan/

Rossi PH, Freeman HE, Lipsey MW (1999), *Evaluation: A Systematic Approach. 6th ed*. New York: Sage Publications.

Rubenstein LV, Mittman BS, Yano EM, Mulrow CD (2000). From understanding health care provider behavior to improving health care: The QUERI framework for quality improvement, *Medical Care* 38(6) QUERI Supplement:I129–I141.

Saultz JW (2003). Defining and measuring interpersonal continuity of care, *Ann Fam Med* 1(3):134–143. As of April 25, 2007:
http://www.annfammed.org/

Saultz JW, Lochner J (2005), Interpersonal continuity of care and care outcomes: A critical review, *Ann Fam Med* 3(2):159–166. As of April 25, 2007:
http://www.annfammed.org/

SGEGL—*See* Air Force Surgeon General's Executive Global Look.

Skinner RI (2003). The value of information technology in healthcare, *Front Health Serv Manage* 19(3):3–15.

Theater Medical Information Program-Joint (TMIP-J) Web site. As of November 9, 2007:
http://www.ha.osd.mil/peo/tmip/default.asp

Thomson Healthcare, Medical Episode Grouper (2007). As of November 13, 2007:
http://home.thomsonhealthcare.com/Products/view/?id=72

TRICARE (2007). *Evaluation of the TRICARE Program: FY 2007 Report to Congress*. Washington, D.C.: Health Program Analysis and Evaluation Directorate, TRICARE Management Activity (TMA/HPA&E), Office of the Assistant Secretary of Defense (Health Affairs) (OASD/HA). As of July 2, 2007:
http://www.tricare.mil/ocfo/_docs/Evaluation_of_the_TRICARE_Program.pdf